DRINKING WATER SUPPLIES

SUPPLIES

Protection through Watershed Management

DRINKING WATER SUPPLIES
Protection through Watershed Management

by

Raymond J. Burby
Edward J. Kaiser
Todd L. Miller
David H. Moreau

ANN ARBOR SCIENCE
THE BUTTERWORTH GROUP

PREFACE

This volume describes a new methodology for devising programs to protect surface water supply sources. Urbanization of water supply watersheds is creating an increasingly serious hazard to public health. In the past, protection of water supply sources relied on the concept of source isolation through purchase of the surrounding area. This practice is no longer practical—on economic, legal, or political grounds—particularly in areas where urban development pressures are accelerating. As a result, water supply systems and local governments must devise watershed management strategies that incorporate a variety of measures and institutional arrangements to protect raw water supplies from contamination.

THE MANAGEMENT PROCESS

In this book, watershed management is proposed as a way to reduce the exposure of water supplies to polluting land uses. Because it is difficult to mandate exactly how privately owned land may be used, watershed management programs must rely on a carefully thought out management process and a broad range of tactics to bring about and maintain proper types of watershed development. In addition, it is important to realize that watershed management is as much an art as a science. In spite of extensive study and the wealth of experience that has accumulated, precise methods of easily assessing land use effects on water quality do not exist. Predicting the health effects of drinking treated water from a polluted source is even more uncertain.

For these reasons, formulating watershed management programs to protect drinking water quality is a complex endeavor. It helps to have a general knowledge of experiences with watershed management in other communities in order to select management approaches of proven effectiveness and to find ways to avoid management obstacles that are typically encountered. In addition, it is vital to have a basic understanding of the inherent uncertainties involved in watershed management to avoid developing a program that is not scientifically creditable. Once this background has been acquired, a community is ready to investigate its water supply protection needs.

Assessing the problems that a watershed management program should address is not an automatic, technical exercise. As potential problems are detected through sanitary surveys and other means, judgments must be made about whether a problem is important enough to warrant further study. Careful investigation of existing and potential land use activities in the watershed, and their potential for pollution, is one of the best means of assessing risks development poses for drinking water quality.

Once threats to water supplies have been determined, a variety of methods may be used to manage watersheds. Land use planning measures are designed to reduce the susceptibility to harm from pollution by affecting the type and location of human activities occurring in the watershed. In addition to the plan itself as a guide to action, regulations (zoning, subdivision regulations, etc.), capital improvements, land acquisition, and preferential taxation can be used to guide urban growth. After it is determined which types of land uses will be allowed in water supply watersheds, a second critical aspect of watershed management programs is their control over the characteristics of site development allowed in the watershed so as to minimize the generation and transport of various pollutants. These site-level pollution control requirements are also applicable to existing land uses which may already be causing water quality problems. Before any given combination of land use and site-level measures is adopted, it should be evaluated in terms of a policy framework that sets overall agency goals; desired equity, efficiency, and feasibility characteristics of the management program; and principles and standards for development and use of the watershed.

Finally, it is important to continuously reassess management efforts to ensure that they are having their intended effects. This is especially important since water quality is still not well understood and new information is frequently becoming available. In addition, changes in political attitudes toward regulating land uses must be closely observed to determine if a different strategy for managing watersheds should be investigated.

The methods for watershed planning covered in this guidebook have been developed in light of two basic principles. First, the problems occurring in an individual watershed are likely to be specific to that

vi

drainage area, depending on the amount, timing, and location of development activities and their relationship to natural features in the watershed. Thus, watershed management strategies must be devised on a case-by-case basis following realistic assessments of probable development patterns and related water quality effects that will occur in the absence of active protection measures. Second, management strategies and management institutions cannot be designed independently of each other. Legal, financial, and political powers and management expertise of government agencies are important determinants of the effectiveness with which alternative strategies may be expected to resolve or prevent water quality problems. Therefore, they should be carefully considered to determine if existing governmental institutions are adequate for the types of management tools that need to be employed in mitigating existing and potential pollution sources threatening a community's watersheds. If it is found that the fit is poor, then either existing institutions need to be modified or different management measures must be employed.

THE GUIDEBOOK

This book is intended to serve five major functions:

1. provide an overview of why watershed management is important and how it is currently practiced throughout the United States;
2. provide an appreciation for the scientific basis for watershed planning, including an understanding of how to deal with the high degree of uncertainty involved in predicting land use effects on water quality;
3. provide technical guidance—what to look at, what to look for, and how to do it—in undertaking a water supply protection program;
4. provide an efficient, scientifically valid, and practical planning process for maintaining the quality of drinking water supplies; and
5. provide perspectives on how to judge the success of water supply management efforts once they have been initiated.

The guidebook is designed to serve water systems and government planning agencies at the local and regional levels, meeting their particular needs and technical capabilities. It is not intended to replace detailed technical information about particular aspects of watershed protection, but rather to act as a resource document to help make sense out of the confusing hodgepodge of reports that have been published about water quality management. Numerous references cited throughout the guidebook provide sources of more detailed information should it be needed.

Three important features to this guide ensure its utility and deserve special mention:

1. It is up-to-date. National surveys of regional and local planning agencies and water system managers provide the most current information that is available on the status of watershed management programs throughout the United States.
2. It is based on everyday experience. In-depth case studies conducted in communities with especially interesting approaches to watershed management provide real-life examples of ways watershed management programs have succeeded and failed.
3. It is based on previous research. A comprehensive review of water quality literature, including technical studies and management plans prepared on the federal, state, and local levels, were used extensively in preparing this guide.

Many more planning methods are discussed in this book than any one community would ever use; however, knowledge of their existence should enable the selection of management methods that are most appropriate, given particular needs. At the outset of a planning program, and throughout the management effort, decisions about the nature and structure of the planning process to be pursued have to be made. These decisions are influenced by the type and extent of water quality problems that exist or threaten, as well as the amount of resources a community is able to devote to water quality protection. Each chapter in this guidebook should assist localities in making judgments about how to proceed with watershed management.

Chapter I reviews the major problems addressed by watershed management and describes the historical evolution of source protection in the United States. The various types and characteristics of pollution found in surface waters are explained briefly. Sources of pollution are identified, impacts on receiving waters are noted, and ultimate health effects are discussed. The chapter concludes by noting that while water treatment once seemed sufficient to deal with problems of water source contamination, total reliance on treatment is no longer acceptable.

Chapter II looks at the state of watershed management practices throughout the United States. In addition to a general literature review, it is based on national surveys of regional agencies, local governments, and water systems that were undertaken to explore what has been done and the problems that have been encountered in protecting public water supplies.

Chapter III outlines a method of assessing the problems a watershed management program should address. It proposes a 12-step procedure for identifying problems that have arisen and will occur in the future if existing management policies and practices are continued. In addition, methods for evaluating the relative importance of problems as a basis for formulating management options are discussed. The chapter also provides guides for making judgments about particular aspects of water quality protection when there is insufficient information to be completely certain about what course of action to take.

The next three chapters focus on various aspects of formulating and implementing a watershed management program. Chapter IV examines the basic, direction-setting dimensions of a watershed management program, including formulating goals and objectives, selecting target pollutants, sources, locations, and hydrologic processes, and specifying appropriate combinations of land uses and land use practices for the watershed. These uses and practices comprise the locational and engineering measures undertaken by agricultural and urban activities, including development practices. Chapter V outlines a range of intervention measures that might be applied in a watershed management program. Specific measures are organized into categories, based on their role in a management program, the nature of governmental powers used, and the basic dimensions of the watershed management program to which the measures are addressed. Chapter VI then addresses the formulation of the actual watershed management program, discussing the major issues that must be addressed and analytic methods available for choosing among alternative courses of action.

Chapter VII concludes this book by suggesting methods for monitoring and evaluating watershed management programs after they are in operation and for making mid-course corrections to improve program performance. The methods described are critical, because with the level of uncertainty inherent in watershed management, adjustments in program design and implementation are almost certain to be required.

This book is a joint product of the four principal authors. Dr. Raymond J. Burby assumed primary responsibility for Chapters I and II and contributed to Chapters VI and VII. Dr. Edward J. Kaiser assumed primary responsibility for Chapters IV and V and contributed to Chapters III and VI. Mr. Todd L. Miller assumed primary responsibility for chapters VI and VII and contributed to Chapters I and V. Dr. David H. Moreau assumed primary responsibility for Chapter III and contributed to Chapter I.

We would like to acknowledge the assistance of Asta C. Cooper and Janice Higgins, who served as research assistants on this project, and Carroll Carrozza and Barbara Rodgers, who provided project administrative and secretarial assistance. We owe a debt of gratitude to the water system and local government officials in our eight case study communities and to the more than 1300 water system managers and local and regional officials from across the country who responded to our requests for information. A final note of thanks is due William Norris, Watershed Management Official in Albemarle County, Virginia, for critically reviewing the draft manuscript.

The work on which this publication is based was supported in part by funds provided by the Office of Water Research and Technology, U.S. Department of the Interior, Washington, DC, as authorized by the Water Research and Development Act of 1978. Any opinions, findings, conclusions, or recommendations expressed herein are those of the authors and do not necessarily reflect the views of the Department of the Interior.

Raymond J. Burby
Edward J. Kaiser
Todd L. Miller
David H. Moreau

Burby　　　　　**Kaiser**　　　　　**Miller**　　　　　**Moreau**

Raymond J. Burby is Assistant Director for Research at the Center for Urban and Regional Studies of the University of North Carolina at Chapel Hill. He received his AB in Government from George Washington University, and his MRP and PhD in Planning from the University of North Carolina. Dr. Burby has had extensive experience in research focusing on environmental planning and management, serving as the principal or co-principal investigator on some 25 sponsored research projects over the past 15 years. He is the author of six books and numerous journal articles, including recent works dealing with energy conservation, flood hazard management, and the use of land use planning techniques in public health programs. Dr. Burby is book review co-editor for the *Journal of the American Planning Association* (APA) and has served as president of the North Carolina Chapter of APA and the North Carolina Land Use Congress, Inc. He currently serves on the Executive Council of the Southern Regional Science Association.

Edward J. Kaiser is Professor of Planning at the University of North Carolina at Chapel Hill, where he teaches planning methods and land use planning in the Department of City and Regional Planning. He received his undergraduate training in Architecture at the Illinois Institute of Technology and his PhD in Planning from the University of North Carolina. His research currently focuses on land use planning for water supply protection, flood hazard mitigation, and energy conservation. He is the co-author of numerous articles and books on planning and the residential development process. Professor Kaiser has been the principal or co-principal investigator on ten major research projects. With Raymond Burby, he is book review editor of the *Journal of the American Planning Association.* He is a past officer of the Association of Collegiate Schools of Planning.

Todd L. Miller, Planning Consultant with the firm Marine Chemurgics, Research and Development, (Ocean) Newport, North Carolina, specializes in coastal and water resources management, land use planning and scientific survey methods. He received his MRP in City and Regional Planning from the University of North Carolina at Chapel Hill. During the past several years he has worked as a Research Associate at the University of North Carolina, specializing in the application of land use planning techniques to water resources problems, and as a consultant for the Washington, DC-based Conservation Foundation. He is president of the North Carolina Coastal Federation, Inc., a nonprofit public interest organization established to improve the management of coastal resources in North Carolina.

David H. Moreau, Professor of Water Resources Planning in the Departments of City and Regional Planning and Environmental Sciences and Engineering at the University of North Carolina at Chapel Hill, received his PhD in Water Resources from Harvard University. In addition to his various research activities in urban water planning and management, he has been a consultant to the U.S. Environmental Protection Agency and local governments on environmental planning and risk assessment. For the past five years he has served on the board of directors of the Orange Water and Sewer Authority, including one term as chairman. He was recently appointed as director of the Water Resources Research Institute of the University of North Carolina.

CONTENTS

FIGURES

TABLES

CHAPTER I

THE POLLUTION PROBLEM AND EVOLUTION OF SOURCE PROTECTION

Passage of the Safe Drinking Water Act of 1974 signalled a renewal of national concern for the quality of public drinking water. Renewed interest in public water supply has been accompanied by new concern for water supply source protection, particularly in areas such as urbanizing watersheds where rapid land use changes are taking place. Although the technology exists to purify contaminated water so that it can be consumed by humans, total reliance upon technological solutions appears to be unacceptable for a number of reasons, including lack of technological sophistication among many of the nation's 30,000 to 40,000 municipal water supply systems, high costs of using advanced treatment methods, and potential health hazards from new synthetic organic chemicals. In recognition of the limitations of drinking water standards and treatment technologies, every major authority stresses the need for source protection as a first line of defense against the contamination of potable water.

In this introductory chapter, we briefly review the major problems addressed by watershed management and describe the historical evolution of source protection efforts in the U.S. In describing the problem—pollution of raw water sources—the various types and characteristics of pollutants commonly found in surface waters are explained. In addition, sources of pollution and calculated rates at which contamination occurs, its long-range impacts on receiving waters, and ultimate health consequences are discussed. From this review, it is clear that watershed management must cope with a number of uncertainties. The chapter concludes by describing the evolution of methods for dealing with this uncertainty, beginning with the total isolation of raw water sources, through a period of heavy reliance on water treatment, to the current system of treatment combined with pre- and post-generation pollution control measures.

POLLUTANT CHARACTERISTICS

Physical and chemical properties of pollutants determine
their path and mode of travel in the environment. Pollutants
can be transported or dissolved by water, absorbed by soil
particles or occur in solid forms. The National Academy of
Sciences Committee on Safe Drinking Water has identified five
classes of materials which can be considered drinking water
contaminants: (1) microbiological life, (2) solid particles
in suspension, (3) inorganic solutes, (4) organic solutes,
and (5) radioactivity.[1] Each of these contaminants, their
characteristics, and possible effects on humans are discussed
in this section.

Microbiological Contaminants

Microbiological contaminants in water include bacteria,
viruses, protozoa, fungi, and algae. Bacteria, viruses and
pathogenic protozoa in drinking water can and do cause ill
health. Since 1971, there has been a noticeable increase in
infectious waterborne diseases in the United States.
Possible reasons for this increase include: (1) improved
reporting of disease outbreaks; (2) an overload on water
treatment plants with sources of water of increasingly lower
quality; and (3) water treatment processes that do not remove
viruses. One problem with water treatment is that its effect
upon viruses such as infectious hepatitis is uncertain. In
addition, breakdowns of water treatment equipment are
continually a threat. Fungi and algae do not appear to be
major causes of waterborne disease, although they can produce
unpleasant tastes and odors in water and are expensive and
difficult to treat.

Particulates

Particles which do not dissolve in water may be organic
or inorganic. Such insoluble particles are derived from
soils and rocks, but they may also result from human
manufacturing processes or natural processes. These contaminants include clays, fibrous particles of asbestos materials,
and organic particles from decomposition of plants and animal
debris in soil. Although clay and other natural organic
particles may not be dangerous in themselves, these solids
are carriers of chemicals, nutrients, and heavy metals.
Furthermore, sedimentation in water supply lakes reduces
their storage capacity, an occurrence which may cause future
water shortages even without population growth. The ability
of sediment (an insoluble particle) to absorb prevalent

2

pollutant in all surface waters, has led to its designation as a major pollutant in the United States.

Inorganic Solutes

Inorganic solutes refer to heavy metals and other inorganic elements dissolved in drinking water. Heavy metals include: barium, cadmium, chromium, lead, mercury, silver, beryllium, cobalt, copper, magnesium, molybdenum, nickel, tin, vanadium, zinc, and sodium. Other inorganic substances in water are arsenic, selenium, fluoride, nitrate, and sulfate.

There is increasing concern about the contamination of receiving waters with heavy metals. They tend to precipitate out of solution with relatively neutral pH value, and they may be absorbed on clay particles or bound by such compounds as iron or manganese oxides. As a result, though the water itself may contain only small amounts of these materials, the particulate matter in water, and especially the benthic deposits, may be significantly polluted.[2]

Health concerns associated with heavy metals are related to the total intake of inorganics from water, air, and food. The adverse effects may be experienced immediately, or they may result from cumulative effects of long-term exposures. Certain substances, such as barium, lead, and nitrates, are most hazardous to children. Other metals, such as cadmium and copper, impose dangers to people who are already suffering from other types of health disorders.[3]

Organic Solutes

Organic contaminants include: (1) pesticides; (2) a variety of organic compounds such as vinyl chloride, nicotine, chloroform, carbon tetrachloride, and benzene; and (3) nutrients such as nitrogen and phosphorous. Although organics are often measured in drinking water supplies, only a fraction of the exact chemical species have been identified. The chemical revolution that exploded following World War II has led to new synthetic organic chemical compounds being developed and introduced almost daily into industry, agriculture, the home, and ultimately the aquatic environment. More than 423 organic chemicals have been identified in fresh water, 325 of which have been found in treated water.[4] Carcinogenic effects of these organic compounds in drinking water pose serious threats to human health. Another cancer hazard stems from the existence of high levels of chloroform in many water supplies which contain large amounts of organics.

Conventional treatment processes are only partially effective in removing organic contaminants from water, and even advanced processes such as activated carbon absorption provide no assurance that all trace chemicals will be removed. In addition, technology for the continuous monitoring of these chemicals is not yet available, so that a determination of their presence and removal is difficult to make.[5]

Nutrients, normally nitrogen and phosphorous, also cause water quality problems. Excessive nutrient input into lakes causes eutrophication. While these compounds do not appear to have any serious health effects, they may cause taste and odor problems in water supplies. Even more important, eutrophication is a strong indication that other, more serious types of pollution, such as sediment, pesticides or other hazardous compounds, are present in the water supply.

Radioactivity

Minute traces of radioactivity are common in all drinking water supplies. Questions concerning the health effects of radiation revolve around the consequences of ingestion of various types of radiation in drinking water in small doses over a long time. Radiation can produce three types of health disorders: damages (1) to genes, (2) to cells, or (3) to offspring.

FACTORS INFLUENCING WATER QUALITY

Pollutants that threaten public drinking water supplies may be introduced by many different types of land use activities and follow a variety of pathways into the aquatic environment. Figure I-1 illustrates three major factors that determine drinking water quality: (1) natural systems of watersheds; (2) land use activities in watersheds and the pollutants they generate; and (3) water treatment. Public health has traditionally been protected from the adverse effects of pollution by water treatment. However, pre- and post-generation pollution controls, coordinated by a watershed management program, have also been devised to eliminate various pollutants before they reach the treatment plant.

4

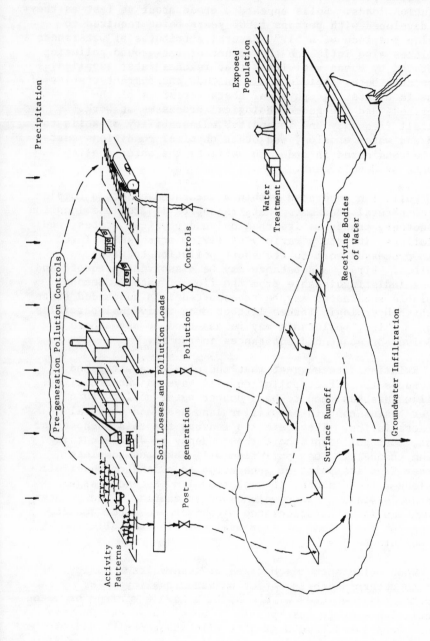

Figure I-1. Pathways Pollutants Take into Public Water Supplies

Water pollution is occurring even before watersheds are disturbed by human activity. Land surfaces erode by normal geological processes, causing what is known as background pollution loads. Soils appear to erode about as fast as they are developed with perhaps 1,000 years being required to develop one inch of soil.[6] Natural attributes of watersheds and lakes also influence the amount of background pollution generated by human activities that reaches water supplies. These attributes include the magnitude and characteristics of climatic events, the physical characteristics of the watershed, and the geomorphological processes at work.[7] Rainfall frequency and intensity, vulnerability of soils to wind and water erosion, and other physical conditions must all be considered in order to estimate the water quality impacts of development.

Pollution generated within a watershed enters a complex web of natural processes that transports it from one location to another, tranforms its chemical and physical states, and assimilates it into a variety of living organisms. Three basic processes work to transport pollution to water supplies. First, a substance may be dissolved in runoff and become indistinguishable from the flow of water. Second, insoluble substances may be transported in a suspended state in which hydrologic forces lift or drag individual particles along. Third, pollutants may be absorbed or may adhere to dissolved or suspended substances in runoff.

Existing data suggest that the size of a watershed influences the effect pollution will have on water quality. Small watersheds tend to have poorer water quality than larger basins containing similar land uses and physical characteristics. There are two reasons for this phenomenon. First, most pollutants have the tendency to settle out in stream channels. Less pollution will make the complete journey from its point of generation to a lake or reservoir the longer the distance it must travel.[8] Second, larger watersheds yield greater volumes of streamflow. The greater streamflow is often associated with improved water quality because of the volume of flow available for the dilution of the pollutants.[9]

When pollutants reach lakes or reservoirs a variety of factors determine their effect on water quality. For example, shallow lakes with many bays have a minimum of water circulation and are susceptible to accelerated eutrophication. Deep lakes have a greater dilution capacity and may be better able to withstand increases in nutrients. Water flow

in and out of the lake (residence time) also affects
vulnerability.[10]

Land Use and Pollution

Pollutants that reach the aquatic environment are
introduced by three major sources: (1) domestic and
industrial wastewater; (2) stormwater runoff; and (3)
accidents. Wastewater discharges are commonly referred to as
point sources of pollution, since they flow out of the end of
a pipe. Stormwater runoff is called nonpoint pollution
because its sources of pollution are widely distributed and
hard to identify. Finally, accidental spills of pollutants
are extremely random and very difficult to forecast.

Point sources are related primarily to domestic and
industrial uses of water. Each person in the United States
uses approximately 75 gallons of water per day for domestic
purposes, a large fraction of which is returned to municipal
sewage treatment systems.[11] Estimates of total industrial
wastewater generated are difficult to make, but they are in
the range of 300 gallons per capita per day.[12] The number of
point sources in any given watershed are normally low enough
that they can be treated individually.

Only fragmentary data are currently available describing
the contributions of nonpoint sources of water pollution.
It is known that as the volume of runoff increases with
greater intensities of land use, the quality of surface
waters decreases.[13] However, stormwater characteristics are
highly variable, exhibiting changes of 10 to 1 or more in
pollutant content during a single storm from area to area and
from one storm to the next.[14] The amount of nonpoint
pollution reaching surface waters is related to runoff
levels. For example, impervious surfaces result in immediate
discharges of rainfall whereas on natural terrain most of it
is absorbed. Runoff coefficients for various land uses have
been estimated based on field studies in a number of areas.[15]
Coefficients are: (1) 10 to 20 percent of rainfall in
forested areas; (2) 50 to 60 percent of rainfall in cultivated areas; (3) 40 to 50 percent of rainfall in residential
areas; and (4) 90 to 100 percent of rainfall in center city
areas. These percentages give a clear indication of the
likelihood of nonpoint pollution with different types of land
use. Even with the difficulty in determining exactly what
types and quantities of pollutants are present in stormwater
runoff, evidence gathered does indicate that urban nonpoint
pollution is significant. Types of pollutants found in
stormwater include microbiological life, sediments, organic

and inorganic solutes, and radioactivity.[16] Obviously, nonpoint pollution constitutes a serious threat to drinking water quality, particularly because it is difficult to divert from water supply lakes and reservoirs.

Although the actual origins of nonpoint pollutants are not always certain, rural and urban development patterns are known to contribute various types and amounts of contaminants. Based on observations made at locations across the country, some estimates have been made about the pollution loads generated when land is used for different purposes. Some of these land uses and their associated loadings are discussed in the following paragraphs.

Rural land uses contribute a significant amount of nonpoint pollution. Sediment and pesticides are the major components of this runoff, although microbiological contaminants may also constitute severe water quality problems. Agricultural erosion produces on the order of 300 to 10,000 tons/mile^2/year of sediment. Cultivated land yields about 400 tons/mile^2/year as compared to pasture land at 325 tons/mile^2/year. Forested lands are estimated to contribute approximately 65 tons/mile^2/year.[17] Chemicals applied to agricultural lands are absorbed by sediment and transported to surface waters. Livestock operations degrade water quality due to over-grazing which results in soil erosion and microbiological contamination from large concentrations of animal feces.[18]

Rural roads are another major source of nonpoint pollution. They produce large amounts of sediment, especially from unpaved portions. It is thought that rural roads especially deserve critical examination because they are generally characterized by less sophisticated engineering and inadequate soil stabilization following construction and periodic maintenance. Moreover, the unpaved roads are particularly vulnerable to accelerated erosion of the roadbed itself as well as its banks and ditches.[19]

Urban land is another major source of nonpoint pollutants. The extent to which land development degrades water quality is dependent on the amount and intensity of urbanization. Results of a recent study of the entire Potomac River Basin indicate just how large an impact urban development can have on the aquatic environment. This study found that the Washington Metropolitan Area produces about 25 percent of the sediment in the river while occupying only 2 percent of the basin.[20]

Sources of water pollution from urban areas include: (1) motor vehicles, (2) construction activities, (3) sewage treatment systems, and (4) recreational, residential, commercial, and industrial land uses and associated activities. Data from recent studies show that contaminants from impervious surfaces are by far the predominant pollutant in urban stormwater runoff.[21]

Nationwide, America's 4 million miles of roads contribute in excess of 56 million tons of sediment per year, or a fourth of all sediment losses.[22] This large volume illustrates the magnitude of water quality problems posed by motor vehicles. In urban areas, strong positive relationships have been found between lead, zinc, and copper and the amount of traffic and percent of impervious cover. This implies that motor vehicles are a major source of metals, and that impervious surfaces deliver them efficiently to storm drainage systems.[23]

The major constituents of street runoff have been found to be suspended and settleable solids--primarily inorganic, mineral-like matter, similar to sand and silt. This material is accompanied by organic matter, algae, nutrients, coliform bacteria, heavy metals, and pesticides. Significantly, the latter are largely concentrated in the very fine materials, which limit the pollution abatement effectiveness of conventional street cleaning equipment or catchment basins.[24] In addition to being transported by runoff, many of these pollutants become airborne and may settle in water supply reservoirs.[25]

Construction-related water pollution is determined by site conditions. The significant variables include size and location of project, rainfall frequency and intensity, pest control measures utilized, resistance of the soil to wind and water erosion, climatic factors affecting the rate and type of revegetation possible, physical and chemical properties of subsurface earth materials, distance to waterways, nature and volume of construction materials used, and requirements for in-water construction operations.[26] Sediment is the chief contaminant generated by construction activities. Without control measures, the loading rate of sediment during construction can be from 10 to 100 times as great on a per-acre basis as the loading rate before development.[27] Sediment yields from small urban areas undergoing construction range from 1,000 to 100,000 tons/mile^2/year.[28]

On-site sewage treatment systems constitute another serious urban threat to water quality. Septic tanks are

9

frequently pinpointed as a major cause of water pollution. Septic tanks are a temporary means of waste disposal, and it is only a matter of time before the waste assimilative capacity of a soil becomes depleted or exhausted. A high percentage of septic tanks fail after a short life period. In all septic tank installations, there is a gradual reduction in the permeability of soils. This results from clogging of soil pores by biological materials and dispersion of soil aggregated by microorganisms. Package treatment plants, a potential alternative to septic tanks, cause problems because they will bypass raw sewage when they are overloaded or not working properly. Sewer and pipeline leakages can also cause considerable problems, and therefore may also pose threats to water quality.[29]

Different forms of urban land uses have varying impacts on water quality. Statistical analysis of a large sample of watersheds indicates that the most important consideration concerning nonpoint pollution loads is the type and intensity of development.[30] As runoff in an urban area increases as a result of development, more contaminants are transported to surface waters. The volume of runoff from urban development is related to types of land uses. As noted above, a much higher proportion of rainfall runs off commercial and industrial land uses than off residential land uses.

Accidental spills constitute another threat to water quality. They generally occur along highways or at industrial or commercial sites. The hazards posed by accidents have increased in recent years because of the greater volume of chemicals and other highly toxic materials being handled by industry or shipped by rail or truck.

Water Treatment

The level of treatment water receives before it is piped to consumers is a third determinant of the quality of drinking water. When water is taken from surface supplies for drinking purposes it normally undergoes a series of broad-band physical and chemical treatments to remove or disinfect contaminants. Coagulation, flocculation, sedimentation, and filtration are treatment processes that remove suspended and some dissolved substances. Chlorination, the usual treatment to disinfect water, kills a wide range of bacteria. More advanced treatment processes, such as activated carbon, are capable of removing some of the toxic substances that may be polluting water supplies.

10

Even more advanced technology exists to purify contaminated water so that it can be consumed by humans. However, for a number of reasons reliance upon these technological solutions is risky. Many public water supply systems are incapable of utilizing more sophisticated water treatment methods needed to deal with various toxic substances.[31] The cost of treatment in many instances represents a considerable burden on the systems' financial capacity.[32/33] With new toxic substances constantly being introduced, drinking water standards and treatment measures are frequently outdated. Drinking water standards often do not adequately take into account antagonistic and synergistic effects among pollutants. Finally, the long-term and threshold effects on health of many potentially carcinogenic substances are not yet known.

Determining if more sophisticated types of water treatment are needed is often difficult for water systems to determine. Water quality monitoring for some contaminants is performed routinely in treating water. Measurements for hard to detect chemicals are made less frequently or are made only during special studies. As a result, unknown chemicals may be present in raw water and they may pass through treatment processes without being removed. They may also be altered to a more harmful state. For instance, carcinogenic trihalomethanes are generated when chlorine reacts with organic materials in raw water supplies. Thus, while public drinking water in the United States is generally regarded as safe, treatment technology alone cannot guarantee that water will not produce harmful effects to public health under all circumstances.

PUBLIC HEALTH AND WATER QUALITY

Consumption of polluted water may result in acute or chronic health effects. Acute effects occur suddenly, soon after a person is exposed to microbiological contamination or toxic levels of chemicals. In most cases, it is possible to recover from the acute effects of drinking polluted water if medical treatment is received promptly. On the other hand, chronic effects are frequently irreversible. They are normally associated with toxic compounds that accumulate in the body over a long period of time, eventually reaching concentrations that will cause cancer or other types of illness. It may take twenty to thirty years for chronic effects of drinking contaminated water to occur.

To date there is no well-established method for assessing risks resulting from land uses in water supply watersheds. The Safe Drinking Water Committee[*] and the U.S. Environmental Protection Agency (EPA)[34] have advanced the furthest in predicting health consequences of drinking polluted water supplies. Both groups have tried to estimate the rates of incidence of disease resulting from the ingestion of water containing chemicals, bacterial populations, and other substances at known concentrations. However, neither has addressed the problem of assessing the health risks to which a population in a given area would be subject under alternative patterns of activity in a watershed.

The Safe Drinking Water Committee and EPA analyses of acute health effects followed classical methods of toxicology to establish "no-observed-adverse-effect" standards. All standards were based on long-term feeding studies on laboratory animals. While the Safe Drinking Water Committee would have preferred to use epidemiological studies, it noted a number of problems that precluded their use as the primary basis for setting standards. Those problems included: (1) limitations on the use of hazardous experiments with human subjects; (2) the absence of an unexposed control group; (3) difficulties in estimating dose levels; (4) difficulties in detecting small changes in effects; and (5) an inability to control for interactions with other pollutants. Epidemiological studies, when available, were used either to confirm animal tests or to detect errors in their use.

Procedures for assessing chronic effects, particularly cancer, are much less developed than those used to study acute effects. There is wide agreement that methods now used to quantify the estimated human risks from a given exposure to a potential carcinogen can provide only rough approximations of the actual risks. Furthermore, it is also believed that methods do not exist for determining a "safe" threshold level of exposure to carcinogens even though such standards have been set based on the best available data. One reason that these standards are questioned is that they are based on animal tests conducted to identify chemical substances that may cause cancer in humans. These tests involve exposing animals to large doses of suspect substances, sometimes in a way different from the way in which humans would come in

---

[*]The purpose of the Committee's study, as defined in Section 1112(e) of the Safe Drinking Water Act of 1974, was to provide the scientific basis and recommended maximum contaminant levels upon which the Environmental Protection Agency would set permanent national drinking water standards.[35]

contact with the substance, to determine if a cancer develops.

From the preceding discussion it should be clear that there is much uncertainty involved in predicting the public health consequences of water supply watershed development and use. It is difficult to determine what and how much pollution will be generated by different types of land uses within a watershed. The uncertainties become compounded when attempts are made to estimate the amount of pollution that is actually delivered to surface water bodies once it is released into the environment. Moreover, questions remain about what happens to pollutants once they reach a water-course and whether they are removed by water treatment processes. Finally, the actual responses of humans who drink polluted water is difficult to predict. Thus, estimates of health hazards from watershed uses and activity levels are rough approximations at best.

In extreme cases of pollution it is possible to prove that significant risks exist from drinking contaminated water. However, in many instances the uncertainties inherent in conducting an assessment of health risks make them of little practical value. In light of these constraints, public health officials have evolved guidelines to minimize the likelihood of adverse health effects.

Expert committees have been used to set drinking water quality standards ever since they were first established in 1914. Committees have been used to enhance participation by the scientific community in the compilation and review of basic scientific data, to make inferences from the data, and to determine appropriate safety factors by which to err on the side of caution in setting water quality standards. The Safe Drinking Water Committee is the most pertinent example of this approach.[36] After reviewing the scientific litera-ture and previously established standards, the Committee made recommendations about the average daily intake of pollutants that it felt would result in no observed adverse effects. It then reduced the average daily intake standard in accordance with the level of uncertainty about its carcinogenicity.

The U.S. Public Health Service contends that the quality and protection of the water supply is the first line of defense to avoid health risks. In 1962 the Advisory Com-mittee to the Public Health Service recommended,

> ...the production of water supplies which pose
> no threat to the consumer's health depends on
> continuous protection. Because of human frailties

associated with this protection, priority should be
given to the purest source. Polluted sources should
be used only when pure sources are economically
unavailable....

The U.S. Environmental Protection Agency has adopted the same
rationale in regard to protecting water supply lakes and
reservoirs. It stresses the importance of the sanitary
survey and recommends that frequent surveys be made to locate
and identify health hazards which might exist in the watershed.[37] A hazard is defined as any condition, device, or
practice in the watershed which creates, or may create, a
danger to the health and welfare of the water consumer.
These hazards should be eliminated at their source if
possible.

Another widely practiced approach to ensure good
drinking water quality is to monitor for pollutants and
remove them by waste treatment or water supply treatment
plants. The adequacy of these technological measures is
currently being questioned. Current monitoring and treatment practices are not thought to be capable of detecting and
removing low concentrations of chemicals that can cause
long-term chronic health effects. Estimates of the national
costs of applying the same level of treatment to nonpoint
pollution that point source pollutants are currently receiving are prohibitive, ranging from $253 billion to $600
billion in 1974 dollars.[38] Scientists believe that we are
still a long way from having technology that is economically
feasible for routine monitoring and treatment of chemicals.[39]
For these reasons, public health officials still believe that
protecting raw water supplies from possible sources of
contamination is an essential aspect of water supply management. In the following section, the evolution of source
protection is traced from the mid-nineteenth century to the
present.

EVOLUTION OF WATER SOURCE PROTECTION

The quest for pure, clean drinking water began in
prehistoric times. However, scientific evidence linking
public health to water supplies was not available until the
mid-nineteenth century. In a now famous epidemiological
study of the contaminated Broad Street Pump, Dr. John Snow
demonstrated in 1854 that cholera rates were much higher
among London residents who were drinking water supplied by
the Southwark and Vauxhall Company from a highly polluted
source than they were among residents supplied by the Lambeth
Company from a much purer source of supply. As the

14

association between human contact with raw water and waterborne diseases such as cholera, typhoid, and dysentery became
clearer, cities increasingly looked to sources of drinking
water which were isolated from centers of population and
potential contamination.

As early as 1834, Boston was advised to seek water well
beyond the urbanized area and by 1848 had tapped Long Pond,
which it renamed Cochituate Reservoir.[40] The Croton Aqueduct
serving New York City was approved in 1835 and completed in
1842. Later in the century Newark switched to an upstream
source of raw water and reduced the mortality rate from
typhoid fever from over 100 per 100,000 residents to 20. As
other cities followed suit, the principle of watershed
management and protection as the first line of defense
against drinking water contamination became well accepted.

Source Isolation

Prior to the advent of modern water treatment technology, the only way to ensure that a pure source of water
remained potable and free of contamination was to exercise
strict sanitary control of the watershed. Regulation of
private landowners to minimize pollution is now well
accepted, but prior to the 1920s and 1930s the only practical
means of sanitary control was acquisition.[41] A number of
water systems establishing surface sources of supply during
the latter half of the nineteenth century, before water
treatment technologies were perfected, purchased vast quantities of land to protect water quality. The City of Newark,
for example, acquired over 35,000 acres in northern New
Jersey to protect its Pequannock Watershed. Seattle's municipal watersheds encompass 450 square miles, approximately
one-quarter of the land area of King County. Smaller
municipally owned watersheds dot New England and other areas
of abundant surface water supplies which experienced rapid
growth during the "pre-treatment" era. Typically, these
watersheds were planted in trees (or left in natural forests)
to minimize sedimentation, pines were planted along reservoir
shorelines to prevent the growth of deciduous trees whose
leaves produce phenolics in the water, and public access was
prohibited.[42]

The Water Treatment Alternative

Source isolation was and still is an effective means of
protecting the public from waterborne disease. However,
until the 1970s its use steadily diminished throughout the

15

twentieth century. There are three related reasons. First, and most important, the development of water treatment technology allowed water purveyors to use less than pure sources of raw water without endangering public health.* Second, when standard treatment measures (coagulation, sedimentation, filtration, and disinfection) were applied, it was no longer cost effective to acquire an entire watershed to guard against contamination of raw water. Third, the growing population of the United States produced increasing pressures to use watershed properties for agricultural and urban uses. In combination, these three factors have resulted not only in decreased interest in source isolation, but until recently, in disinterest in watershed management generally. Thus, over time the "first line of defense" became no line at all.

During the latter two decades of the nineteenth century and first decade of the twentieth century, enormous strides were made in water treatment technology. In 1872, a successful slow sand filter was installed in Poughkeepsie, New York, which drew its raw water from the Hudson River. Although it was not until 1893 that the next successful plant was built in Lawrence, Massachusetts, filtration experiments at Lawrence and Louisville, Kentucky during this period showed that it was possible to treat poor quality raw water so that it could be consumed without endangering public health. In addition to eliminating turbidity and color, filtration removed about 99 percent of the bacteria present in water prior to treatment.[43] Filtration reduced the threat of waterborne disease, but it did not eliminate all pathogenic organisms. The next major step forward in water treatment technology occurred in England in 1904, where hypochlorite (bleaching powder) was first introduced to disinfect water before it entered the London distribution system. In the latter part of 1908, a system for treating Jersey City's raw water with dry calcium hypochlorite was installed by the East Jersey Water Company, after which the city's high typhoid fever rate immediately dropped. The use of chlorination to eliminate pathogenic organisms in drinking water spread rapidly across the U.S., so that by the 1940s over four of every five water systems applying any treatment used chlorination.[44]

------

*Faith in water treatment as an effective means of guarding public health is exemplified in a brochure published by the Water and Wastewater Equipment Manufacturers Association in the mid-1960s, "Water treatment methods can produce safe water out of a seriously polluted raw water source." The Association did concede, however, "Of course, the better the source the less expensive the treatment."[45]

Prior to World War I all of the basic elements of modern water treatment were in place. In 1914, the first bacteriological standard for drinking water was adopted by the U.S. Treasury Department.[46] The standard applied only to interstate carriers, but it provided a model for the subsequent adoption of standards by state and local governments and water purveyors. In the 1925 revision of its drinking water standards, the Treasury Department required that water be obtained from either a source free from pollution, or a source adequately protected by natural agencies from the effects of pollution, or a source protected by artificial treatment.[47] Here, for the first time, it was officially recognized that water treatment could substitute for source isolation and protection. In 1934, the American Water Works Association adopted the position that while watershed management provides the first line of defense against contamination, some form of treatment is also necessary. Over the following four decades treatment standards were revised and upgraded (in 1942, 1962, and 1977), while attention to standards and procedures for watershed protection waned. As Okun has noted, "... with the development of water treatment technology, particularly disinfection with chlorine, at the beginning of this century, engineers became sanguine about the dangers of using polluted sources because these could be rendered safe by appropriate treatment."[48]

The Consequences of Neglect

As new water systems were established in the years after World War I and existing systems were expanded, source isolation fell into general disuse. Water supply utilities did not acquire or control watersheds feeding their reservoirs and often allowed some uses, such as recreation, of the reservoir shoreline and water surface. While recreational activities, if properly managed, have been shown to be compatible with the protection of public health,* other watershed uses may have serious adverse health consequences. By the end of the 1960s evidence of these problems began to mount. In 1969 the Bureau of Water Hygiene of the Public Health Service conducted a nationwide survey of community water supply systems. Only 59 percent of the systems met all of the Public Health Service drinking water quality standards and 26 percent reported specific problems with surface raw water quality.[49] In all, the Public Health Service indicated that over 25 million persons were obtaining substandard drinking water and that approximately 8 million were

---

*See, for example, reference 50.

consuming "potentially dangerous" water. The two most frequently noted deficiencies in water supply systems were lack of adequate source protection and lack of adequate treatment facilities.

In 1972, the President's Council on Environmental Quality reported that over 90 percent of the nation's watersheds were more than "moderately" polluted. In 1974, while the Safe Drinking Water Act was under consideration by Congress, reports were released indicating that a number of cancer-causing chemicals were present in the drinking water of New Orleans, Jefferson Parish, Evansville, Duluth, and Cincinnati. Expanded tests of drinking water in 113 cities during 1976-1977 turned up at least traces of suspect chemicals in each metropolitan area studied and high levels of known or suspected carcinogens.[51] Finally, the incidence of waterborne diseases which were once thought to be well under control has been increasing gradually since the early 1950s. Recent statistics indicate that from 1961 through 1978 drinking water caused 407 outbreaks of disease or poisoning, resulting in 101,243 recorded illnesses and at least 22 deaths. Some water supply experts believe that ten times as many outbreaks actually occur as are officially reported.[52] Although many of these outbreaks have been traced to inadequate water treatment, recent outbreaks of giardiasis are of particular concern because they have occurred in surface-source water systems where treatment facilities were operating properly.[53]

The Safe Drinking Water Act and Source Protection

The Safe Drinking Water Act of 1974 (P.L. 93-523) was enacted by Congress on December 14, 1974. Stimulated by increased national concern with the pollution of drinking water, the act for the first time gave the federal government the power to regulate drinking water quality in public and private water systems.[*] The act authorized the Environmental Protection Agency (EPA) to establish and enforce national regulations to protect the public from contaminated water. Specifically, EPA was required to specify maximum levels of contaminants in drinking water and to set requirements for laboratory testing, monitoring, record keeping, and reporting systems. National Interim Primary Drinking Water Standards were formulated by EPA and became effective on June 24, 1977.

---

[*]Prior to enactment of the Safe Drinking Water Act, the federal government only regulated drinking water used by interstate carriers.

18

The Safe Drinking Water Act is directed primarily (except for the protection of groundwater) at finished water quality and does not set standards for or otherwise address specifically the issue of surface source protection and raw water quality.* However, as noted above, it is increasingly recognized that it is impractical to rely solely on water treatment to protect public health from contaminants in drinking water. There are four reasons. The Environmental Protection Agency recognizes the limitations of drinking water standards and treatment technologies. The first of three approaches identified by EPA for meeting the Congressional goal of assuring that the public has safe water to drink was "1. To ensure that all underground and surface sources of water which are (or potentially may be) used as drinking water supplies are protected to the extent feasible so that the public water systems using those waters are not prevented from attaining the standards established in the National Primary Drinking Water Regulations."[54] However, moves to achieve this source protection goal were not provided through the Safe Drinking Water Act. Instead, federal efforts to protect raw water quality have been diffused across several major programs, including the Federal Water Pollution Control Act and its amendments, the Toxic Substances Control Act, and the Resource Conservation and Recovery Act. Although water supply watershed protection has been considered in the administration of these programs, federal regulations designed specifically to protect drinking water sources have not been forthcoming, in large part because land use management is still viewed as a state and local rather than a federal responsibility.

Local Government and Drinking Water Quality

Believing public health to be adequately protected by water treatment, prior to 1970 few state or local governments were actively concerned with protecting water supply sources from contamination. In 1970, states were spending about a third of the amount experts estimated they should have spent on supervisory programs to assure safe drinking water.[55] State health codes often referred to the need to protect water sources from contamination, but few states required local sanitary surveys to assure that pollution sources were

---

*Although the Safe Drinking Water Act did not mandate source protection standards, federal water supply guidelines have long stressed the importance of sanitary surveys of water supply watersheds and have linked the degree of treatment required to the quality of source waters.[56]

abated and the codes were actually enforced.[57] At the local
level, water supply, water quality, and land use planning
were treated as if there was no relationship among them.[58]
As noted above, water supply agencies often had little choice
but to use raw water which had been exposed to various
sources of contamination; in fact, Okun has estimated that
about half of the population served by public water supplies
was consuming water from sources which in part consisted of
wastewaters discharged just hours or days earlier from
industrial and municipal sewers upstream.[59] Water quality
agencies tended to be single-purpose, often somewhat autonomous city departments or metropolitan agencies concerned
primarily with the provision and operation of efficient
sewerage systems rather than the general enhancement of water
quality.[60] Local planning agencies, for their part, had
little or no expertise in water resources planning and water
quality management. For all of these reasons, prior to the
1970s local watershed protection programs were infrequently
used to protect drinking water sources from contamination,
and where they were employed they were often ineffective.

For example, a mid-1970s survey of Southeastern states
covered 381 watersheds from which water was drawn to serve 12
million persons. Only 10 percent of the watersheds surveyed
were closed, while another 29 percent were restricted to
certain uses. In over 60 percent of these watersheds no
mechanisms were in place to control potentially polluting
land uses.[61] Reviewing a number of states' and localities'
efforts to control nonpoint sources of lake and reservoir
pollution (sources which are diffused, intermittent, and not
easily contained in sewerage pipes), Berger and Kusler concluded, "...efforts adopted to date have been only partially
successful in dealing with non-point sources of pollution...."[62]. In a similar vein, Lovelace and Cantine summed
up difficulties in managing watersheds which cross juris-
dictional lines by observing, "Land use control measures
watered down enough to receive...support are not likely to be
truly effective," and, more generally, "...local control of
land use in the United States has been... ineffective."[63]

Winds of Change

During the 1970s, a number of factors came together to
produce greatly increased interest and correspondingly
greater water system and local government activity in
protecting drinking water sources from contamination. The
1969 federal community drinking water survey referred to
earlier drew professionals' attention to the widespread
deficiencies existing in water treatment and source

20

protection procedures. Additional problems were revealed by the National Eutrophication Survey, initiated by EPA in 1972, which showed that a number of water supply reservoirs were eutrophic. However, neither this finding nor the earlier community drinking water survey results drew sufficient public attention to induce widespread action. When the Environmental Defense Fund released a study on November 6, 1974 showing a significant association between drinking water from the Mississippi River and cancer mortality rates in New Orleans, the nation did pay attention. Congress responded almost immediately by passing the Safe Drinking Water Act, which had been languishing in committee for four years. The New Orleans findings were sufficiently startling to lead water systems and local governments across the country to take a new look at their own water supply sources and to begin to consider how to protect them from contamination.

During the early 1970s, local planners began to realize that land use decisions can affect water quality. In his pioneering work, Design with Nature, Ian McHarg related the type and density of land development to water quality and showed how to identify land which is susceptible to erosion and flooding and land suitable as aquifer recharge areas.[64] More sophisticated techniques were subsequently developed by Tourbier and Westmacott in a series of studies of the Christina Basin in Delaware[65/66] and by the American Society of Planning Officials in research undertaken for EPA.[67] The institutional means for putting this new knowledge to work in planning was provided by Section 208 of the Federal Water Pollution Control Act Amendments of 1972 (P.L. 92-500). Under Section 208, it was recognized that in many areas non-point sources of pollution are as significant a threat to water quality as point sources such as municipal and industrial wastewater. Section 208 provided funds to regional agencies which allowed them to consider both point and non-point sources of pollution as they formulated water quality plans. It also allowed agencies to develop mechanisms for solving water quality problems using both conventional (sewerage system extensions) and unconventional (land use management) methods to abate existing and future pollution sources. Although EPA guidelines for implementing the 208 program were never very clear regarding drinking water protection,[68] in a number of areas the program was instrumental in alerting local planners to the possibilities of protecting water quality through land use planning and control.*

---

*A number of examples of the role Section 208 plays are described in the companion volume to this Guidebook.[69]

21

Now, after a decade of rapid change in water quality planning and watershed management, it is appropriate to take stock of where we stand in protecting drinking water sources from contamination. A number of questions need to be addressed. For example, how widespread is urban encroachment on water supply watersheds? Are water systems aware that urban encroachment can lead to problems of water yield and raw water quality? What methods are being used to cope with water quality problems...by water systems?...by local governments?...by regional agencies?...by state and federal agencies?... by the private sector? What obstacles have agencies encountered in their efforts to protect drinking water sources? Overall, how successful have watershed management programs been in preserving and enhancing raw drinking water quality? These issues are addressed in the following chapter of this Guidebook.

REFERENCES

1. Safe Drinking Water Committee, National Academy of Sciences, Drinking Water and Health, Washington, D.C., 1977.

2. Hammer, Thomas R. Planning Methodologies for Analysis of Land Use/Water Quality Relationships, Water Planning Branch, Environmental Protection Agency, Washington, D.C., 1976.

3. Rostow, Elspeth. Options for Community Response to Safe Drinking Water Act, Lyndon B. Johnson School of Public Affairs, The University of Texas, Austin, 1979.

4. Okun, Daniel A. "Drinking Water Quality Through Enhancement of Source Protection," in Drinking Water Quality Enhancement Through Source Protection, Robert B. Pojasek, ed., Ann Arbor Science Publishers, Inc., Ann Arbor, Mich., 1977(a).

5. Okun, Daniel A. "Drinking Water Quality Through Enhancement of Source Protection," in Pojasek, 1977(a).

6. Brandt, G.H., E.S. Conjhers, F.J. Lowes, J.W. Mighton, and J.W. Pollack, An Economic Analysis of Erosion and Sediment Control for Watersheds Undergoing Urbanization, Dow Chemical Company, Midland, Mich., 1972.

7. Mullane, Neil J. and Gary L. Beach. Planning for the Proper Use of Land and Water, Water Resources Research Institute, Oregon State University, Corvallis, Or., 1977.

8. Hammer, Thomas R. Planning Methodologies for Analysis of Land Use/Water Quality Relationships, Water Planning Branch, Environmental Protection Agency, Washington, D.C., 1976.

9. Nutter, Wade L. The Relationship of Land Use to Domestic Water Supply in Georgia, University of Georgia, Athens, 1973.

10. Sargent, Frederic O., Philip R. Berke, and E. Bennette Henson. "A Water Resources Planning Procedure for Rural Towns," Water Resources Bulletin, Vol. 15, No. 2 (1979), pp. 496-505.

11. Pavoni, Joseph L., ed. Handbook of Water Quality Management Planning, Van Nostrand Reinhold Company, New York, 1977.

12. Todd, David Keith, ed. The Water Encyclopedia, Water Information Center, Port Washington, N.Y., 1970.

13. Michigan Department of Natural Resources. Inland Lake Watershed Analysis: A Planning and Management Approach, Land Resources Program Division, Ann Arbor, Mich., 1976.

14. Lager, John A. and William G. Smith. Urban Stormwater Management and Technology: An Assessment, National Environmental Research Center, U.S. Environmental Protection Agency, Cincinnati, Oh., 1974.

15. Michigan Department of Natural Resources. Inland Lake Watershed Analysis: A Planning and Management Approach, Land Resources Program Division, Ann Arbor, Mich., 1976.

16. Kuhner, Jochen, Russell deLulia, and Michael Shapiro. "Assessment of Existing Methodologies for Evaluation and Control of Watershed Land Use in Drinking Water Supply Systems," in Pojasek, 1977(a).

17. Brandt, G.H., E.S. Conjhers, F.J. Lowes, J.W. Mighton, and J.W. Pollack. An Economic Analysis of Erosion and Sediment Control for Watersheds Undergoing Urbanization, Dow Chemical Company, Midland, Mich., 1972.

23

18. White, Carleston S. "Factors Influencing Natural Water
    Quality and Changes Resulting from Land-Use Practices,"
    Water, Air, and Soil Pollution, Vol. 6, No. 6 (1976),
    pp. 53-64.

19. Hafley, William L. "Rural Road Systems as a Source of
    Sediment Pollution—A Case Study," Watershed Management,
    American Society of Civil Engineers, New York, 1972.

20. Wildriek, John J. Urban Water Runoff and Water Quality
    Control, Virginia Polytechnic Institute and State
    University, Blacksburg, Va., 1976.

21. Kuhner, Jochen Russell deLulia, and Michael Shapiro.
    "Assessment of Existing Methodologies for Evaluation and
    Control of Watershed Land Use in Drinking Water Supply
    Systems," in Pojasek, 1977(a).

22. Brandt, G.H., E.S. Conjhers, F.J. Lowes, J.W. Mighton,
    and J.W. Pollack. An Economic Analysis of Erosion and
    Sediment Control for Watersheds Undergoing Urbanization,
    Dow Chemical Company, Midland, Mich., 1972.

23. Kim, Jung I. Dennis R. Helsel, Thomas J. Grizzard, and
    Clifford W. Randall. "Land Use Influences on Metals in
    Storm Drainage," Journal of Water Pollution Control,
    Vol. 51, No. 4 (1979), pp. 709-717.

24. Lager, John A. and William G. Smith. Urban Stormwater
    Management and Technology: An Assessment, National
    Environmental Research Center, U.S. Environmental
    Protection Agency, Cincinnati, Oh., 1974.

25. Wildriek, John J. Urban Water Runoff and Water Quality
    Control, Virginia Polytechnic Institute and State
    University, Blacksburg, Va., 1976.

26. Pavoni, Joseph L., ed. Handbook of Water Quality
    Management Planning, Van Nostrand Reinhold Company,
    New York, 1977.

27. Hammer, Thomas R. Planning Methodologies for Analysis
    of Land Use/Water Quality Relationships, Water Planning
    Branch, Environmental Protection Agency, Washington,
    D.C., 1976.

28. Brandt, G.H., E.S. Conjhers, F.J. Lowes, J.W. Mighton,
    and J.W. Pollack. An Economic Analysis of Erosion and
    Sediment Control for Watersheds Undergoing Urbanization,
    Dow Chemical Company, Midland, Mich., 1972.

29. Tourbier, Joachim and Richard Westmacott. _Water Resources Protection Measures in Land Development—A Handbook_, Water Resources Center, University of Delaware, Wilmington, 1974.

30. Coughlin, R.E. an T.R. Hammer. _Stream Quality Preservation Through Planned Urban Development_, U.S. Environmental Protection Agency, Washington, D.C., 1973.

31. Pojasek, Robert B. "How to Protect Drinking Water Sources," _Environmental Science and Technology_, Vol. 11 (April 1977(b)).

32. Moody, Tom. "Local Governments and the Safe Drinking Water Act," in _Safe Drinking Water: Current and Future Problems_, Clifford S. Russell, ed., Resources for the Future, Washington, D.C., 1978.

33. Gerber, Robert G. "Land Use Controls in Watershed and Aquifer Recharge Areas," _Journal of the Maine Water Utilities Association_, Vol. 49, No. 6 (1973), pp. 122-126.

34. 44 _Federal Register_ 15926, May 1979.

35. Safe Drinking Water Committee, National Academy of Sciences, _Drinking Water and Health_, Washington, D.C., 1977.

36. Safe Drinking Water Committee, National Academy of Sciences, _Drinking Water and Health_, Washington, D.C., 1977.

37. U.S. Environmental Protection Agency, _Manual for Evaluating Public Drinking Water Supplies_, U.S. Environmental Protection Agency, Washington, D.C., 1975.

38. Municipal Environmental Research Laboratory, _Areawide Assessment Procedure Manual_, Volume 1, U.S. Environmental Protection Agency, Cincinnati, Oh., 1974.

39. Okun, Daniel A. "Drinking Water Quality Through Enhancement of Source Protection," in Pojasek, 1977(a).

40. Council on Environmental Quality, _Recreation on Water Supply Reservoirs: A Handbook for Increased Use_, U.S. Government Printing Office, Washington, D.C., 1975, p. 8.

41. Kusler, Jon. _Regulating Sensitive Lands_, Ballinger Publishing Company, Cambrige, Mass., 1980, p. 7.

42. Ring, Chester A. III. "The Water Supply Industry and Source Protection," in Pojasek, 1977(a), p. 63.

43. Safe Drinking Water Committee, National Academy of Sciences, _Drinking Water and Health_, Washington, D.C., 1977, pp. 2-3.

44. Council on Environmental Quality, _Recreation on Water Supply Reservoirs: A Handbook for Increased Use_, U.S. Government Printing Office, Washington, D.C., 1975, p. 17.

45. National Water Institute, Water and Wastewater Equipment Manufacturers Association, _Background on Water for Municipal, State and Federal Planners_, The Institute, New York, n.d. (circa 1966), p. 7.

46. U.S. Treasury Department. _Bacteriological Standards for Drinking Water_, Public Health Report, Vol. 29(1914), pp. 2959-2966.

47. _Drinking Water Standards,_ Reprint No. 1029 from the Public Health Reports (April 10, 1925), p. 693.

48. Okun, Daniel A. "Drinking Water Quality Through Enhancement of Source Protection," in Pojasek, 1977(a), p. 319.

49. McCabe, L.J. et al. "Survey of Community Water Systems," _Journal of the American Water Works Association_, Vol. 62 (1970), p. 670.

50. Peavy, Howard S. and Claude E. Matney. "The Effects of Recreation on Water Quality and Treatability," in Pojasek, 1977(a), pp. 461-475.

51. Zwick, David. " Our Inland Waters," in _Life After '80: Environmental Choices We Can Live With_, Brick House Publishing Company, Andover, Mass., 1980.

52. U.S. General Accounting Office. _State's Compliance Lacking in Meeting Safe Drinking Water Regulations_, Document Handling and Information Services Facility, U.S. General Accounting Office, Gaithersburg, Md., March 3, 1982, p. 1.

53. Center for Disease Control. "Waterborne Giardiasis—
    California, Colorado, Oregon, Pennsylvania," Morbidity
    and Mortality Weekly Report, Vol. 29, No. 11 (1980),
    Atlanta.

54. U.S. Environmental Protection Agency. "National Safe
    Drinking Water Strategy (draft)," Environmental
    Protection Agency, Washington, D.C., May 1975.

55. Pojasek, Robert B. "Preface," in Pojasek, 1977(a),
    p. v.

56. U.S. Environmental Protection Agency, Office of Water
    Programs, Water Hygiene Division. Health Guidelines for
    Water and Related Land Resources Planning, Development
    and Management, U.S. Government Printing Office,
    Washington, D.C., October 1971.

57. Castorina, Anthony R. "Surveillance and Monitoring
    Program for Connecticut Public Water Supply Watersheds,"
    in Pojasek, 1977(a), pp. 137-143.

58. Schwartz, Seymour I. Robert A. Johnston, James R.
    Blackmarr, and David E. Hansen. Controlling Land Use for
    Water Management and Urban Growth Management: A Policy
    Analysis, California Water Resources Center, University
    of California, Davis, October 1979, p. 1.

59. Okun, Daniel A. "Drinking Water Quality Through
    Enhancement of Source Protection," in Pojasek, 1977(a),
    p. 319.

60. Kaiser, Edward J. et al. Promoting Environmental Quality
    Through Urban Planning and Controls, Socioeconomic
    Environmental Studies Series, Office of Research and
    Development, U.S. Environmental Protection Agency,
    Washington, D.C., February 1974, p. 223.

61. Dissmeyer, G.E. and W.T. Swank. "Municipal Watershed
    Management Survey," Journal of the American Water Works
    Association, Vol. 68 (February 1976).

62. Berger, Bernard B. and Jon A. Kusler. "Lakeshore Zoning
    to Control Nonpoint Sources of Pollution," in
    Urbanization and Water Quality Control, William Whipple,
    Jr., ed., American Water Resources Association,
    Minneapolis, Minn., 1975, pp. 169-179.

63. Lovelace, Eldridge and T. Scott Cantine. "Management-Institutional Aspects of Water Quality Management Planning," in Pavoni, 1977, pp. 322, 332.

64. McHarg, Ian. Design with Nature, Doubleday, Garden City, N.J., 1969.

65. Tourbier, Joachim et al. The Christina Basin: The Protection of Water Resources as a Basis for Planning in Developing Areas, University of Delaware Water Resources Center, Newark, Del., 1973.

66. Tourbier, Joachim and Richard Westmacott. Water Resources Protection Measures in Land Development—A Handbook, Water Resources Center, University of Delaware, Wilmington, 1974.

67. Thurow, Charles, William Toner, and Duncan Erley. Performance Controls for Sensitive Lands: A Practical Guide for Local Administrators, Office of Research and Development, U.S. Environmental Protection Agency, Washington, D.C., 1975.

68. Goldrosen, John. "The Role of Section 208 Planning in Protecting Drinking Water Sources," in Pojasek, 1977(a), pp. 39-61.

69. Miller, Todd L. and Raymond J. Burby with Edward J. Kaiser and David H. Moreau. Protecting Drinking Water Supplies Through Watershed Management: A Casebook for Devising Local Programs, Center for Urban and Regional Studies, The University of North Carolina at Chapel Hill, Chapel Hill, N.C., August 1981.

CHAPTER II

THE STATE OF PRACTICE

The task of watershed management is now more complex and
more difficult than ever before. It is also more vital to
public health. Many water sources, once protected from
contamination by their isolated, rural location are
threatened by creeping urban and exurban development. Conventional water treatment, once relied upon to protect public
health even when suspect sources of supply were used, is
unable to eliminate many viruses and potentially harmful
chemicals which may find their way into raw water supplies.
Coping with this double-barrelled threat requires a variety
of land and water quality management measures. It also
requires the active involvement of water purveyors and every
level of government. Drawing on the results of three
national surveys, this chapter describes the approaches which
are being used across the country to protect drinking water
supplies and evaluates their effectiveness.

This national overview of watershed management
experience may be used in a variety of ways by water system
managers and local planners. The chapter reports water
systems' actual experiences with watershed problems ranging
from insufficient water yield to deteriorating water quality.
The frequency with which problems have occurred is indicated
and sources of problems are identified. This information can
be used by water systems to anticipate problems which are
likely to accompany changing land uses in the vicinity of
their own water supply sources.

The chapter summarizes measures adopted by private
landowners, water systems, local governments, and regional,
state, and federal agencies to cope with raw water problems.
Various mangement tools are described and water systems' and
local governments' ratings of their effectiveness are
reported. This information can be used by water systems to
identify management approaches and measures which might be
effective in their own local circumstances.

Various obstacles which water systems and local
governments have encountered in formulating and implementing
watershed protection programs are analyzed. This information

29

provides water system managers with a series of firsthand lessons they can use to avoid similar problems in developing their own management programs. Finally, the surveys summarized in this chapter constitute the most comprehensive, up-to-date data base ever assembled for describing watershed management as it is practiced in the United States. This information can serve as a benchmark which water systems and local governments may use to gauge their own progress in protecting surface water supply sources in relation to the state of the art nationwide.

NATIONAL SURVEYS OF COMMUNITY WATER SYSTEMS AND LOCAL GOVERNMENTAL AGENCIES

Data to describe and assess watershed management practices across the country were assembled through three surveys: one directed to water system managers; one to local governmental managers and planners; and one to regional councils of government. Each survey was conducted by mail during the spring of 1981 using techniques perfected by Dillman at Washington State University.[1]

Water Systems Studied

Information was obtained from 496 community water system managers.* Water systems were included in the survey if they met two criteria: (1) use of a surface source for at least part of their raw water supply; and (2) provision of service to between 5,000 and 500,000 persons. The second criterion was employed to screen out the thousands of very small community water systems which lack the financial and personnel resources needed to manage water supply watersheds and the very few extremely large systems whose experiences would not be representative of the majority of water systems in the U.S. to whom this book is addressed. The national sample was drawn using lists supplied by the U.S. Environmental Protection Agency. Ninety-one percent of the water systems surveyed responded and returned a questionnaire.

These water systems are located in all fifty states. Most water systems (90 percent) are publicly owned. Municipal systems (74 percent) predominate, but county (3

---

*Respondents described their positions as follows: water system president, director or manager (39 percent); water system superintendent (43 percent); public service/works director (7 percent); other (11 percent).

percent), city/county (3 percent), and special districts (10 percent) are also represented among the public systems. Among the privately owned water systems, nearly half are located in two of the ten federal regions, Region III (Middle Atlantic states) and Region VIII (Rocky Mountain states). The median water system studied serves 17,000 persons, producing an average of 2.6 million gallons of water per day. About a quarter of the systems serve fewer than 10,000 persons, while another quarter serve 50,000 or more.

As shown in Table II-1, the most common raw water source was a lake, reservoir or pond (used by 59 percent of the water systems). In Federal Regions I, II and V, lakes were used as a raw water source by 80 percent or more of the water systems and provided at least half of the raw water treated by three quarters of these systems. River intakes were employed by about a third of the water systems studied and were the leading source of supply in the Pacific Northwest (Region X). Groundwater was also heavily relied upon in the Far West and also in New England, but since water systems were sampled on the basis of their use of surface sources, groundwater was less frequently used by the water systems studied than in the nation as a whole. Finally, 9 percent of these water systems imported water from other systems.

Over 90 percent of the water systems reported that the finished water produced by their treatment plants met all federal and state drinking water quality standards. Larger water systems (those serving more than 50,000 persons) and systems in the Far West were least likely to be meeting all applicable standards.

Information from Local and Regional Agencies

In addition to the national survey of community water system managers, two other surveys were conducted to collect information for this chapter. Because a number of watershed management measures, such as land use regulations, are implemented by agencies other than water systems, a survey of local governments was undertaken to learn about their perspectives on various management issues. This survey was directed to county officials, since most water supply watersheds are located, at least in part, outside of the corporate limits of towns and cities. Counties were included in the local government survey if they contained the headquarters city of one of the water systems included in the study (see above). Questionnaires were returned by 76 percent (314) of the 413 county officials queried. A majority of the county government respondents were county

31

Table II-1. Source of Drinking Water Supply

Federal Region (States)	Percent of Water Systems Using Source:[a]			
	Ground-water	River Intakes	Lakes, Reservoirs, Ponds	Imported from Other Systems
All Regions	35	36	59	9
I (CT, ME, MA, NH, RI, VT)	60	17	85	9
II (NJ, NY)	41	20	85	13
III (DE, MD, PA, VA, WV)	29	45	71	8
IV (AL, FL, GA, KY, MS, NC, SC, TN)	20	38	65	8
V (IL, IN, MI, MN, OH, WI)	30	28	80	2
VI (AR, LA, NM, OK, TX)	22	20	67	7
VII (IA, KS, MO, NB)	27	47	73	7
VIII (CO, MT, ND, SD, UT, WY)	30	52	59	15
IX (AZ, CA, HA, NV)	52	44	52	20
X (AK, ID, OR, WA)	50	67	37	4

[a]Percentages do not add to 100 percent because some water systems used more than one type of raw water source.

managers and planning directors.

The third national survey was undertaken to help us describe regional agencies' activities in watershed management. As noted previously, the federal water quality planning (208) program has been based largely at the regional level. In addition, because water supply watersheds often encompass more than one local governmental jurisdiction, we felt that regional agencies might be playing critical roles in coordinating local governments' efforts to protect raw water quality. The regional agency survey was directed to each of the 652 member agencies of the National Association of Regional Councils. Responses were obtained from 84 percent (549 agencies), with information most often supplied by the agency director/assistant director or planning director and staff.

URBANIZATION OF WATER SUPPLY WATERSHEDS

In recent years, the trend toward population decentralization has markedly increased, placing correspondingly increased pressures on rural water supply watersheds which were once thought to be safe from contamination. Between 1970 and 1980 the U.S. population increased by 11.4 percent, but the nonmetropolitan population increased by 15.1 percent. Nonmetropolitan counties are absorbing an increasing proportion of national growth, jumping from 9 percent of growth occurring between 1960 and 1970 to 25 percent between 1970 and 1980.[2]

Nationwide, 86 percent of the water systems surveyed reported some development of the watersheds from which they drew raw water supplies. Although urbanization is still light (under one of every ten acres in urban use) in the majority of these watersheds, the proportion of water systems experiencing moderate to heavy development pressures—one of every four or more acres already in urban use—had increased markedly, to 23 percent, by 1981. As shown in Table II-2, development pressures tend to be particularly heavy in New England (Region I), the Southeast (Region IV), and Southwest (Region VI).

Urbanization is a major cause of water quality degradation. As a water supply watershed develops, urban runoff reaching the water supply may contain a variety of pollutants, including suspended sediments and toxic materials such as heavy metals, bacteria, oxygen-demanding materials, nutrients, and oil and grease. Water system managers' perceptions of urbanization as a problem are illustrated in

33

Table II-2. Urbanization of Water Supply Watersheds

Federal Region (States)	Percent of Water Systems with Watersheds:	
	10% or More Developed	25% or More Developed
All Regions	46	23
I (CT, ME, MA, NH, RI, VT)	54	29
II (NJ, NY)	40	26
III (DE, MD, PA, VA, WV)	39	17
IV (AL, FL, GA, KY, MS, NC, SC, TN)	58	26
V (IL, IN, MI, MN, OH, WI)	49	18
VI (AR, LA, NM, OK, TX)	60	34
VII (IA, KS, MO, NB)	40	27
VIII (CO, MT, ND, SD, UT, WY)	32	12
IX (AZ, CA, HA, NV)	37	23
X (AK, ID, OR, WA)	5	0

Figure II-1, which shows how the proportion of water systems reporting urban problems increases with increasing watershed development. With less than 1 percent of a watershed in urban use, only 20 percent of the managers felt that urban encroachment had become a problem. This increased to 41 percent as development increased to between 1 percent and 9 percent of a watershed and to 67 percent of the managers as the proportion of development in the watershed approached 25 percent. These data clearly illustrate the importance of limiting watershed development as much as possible. Methods water systems and local governments are using to achieve this goal are discussed later in this chapter. First, however, we would like to take a closer look at water supply problems associated with watershed urbanization.

PROBLEMS ASSOCIATED WITH URBANIZATION

Water system managers were asked about the specific problems their systems were encountering and about potential problems which might occur over the coming decade. Their responses further illustrate the difficulties water systems experience as watersheds change from rural to urban land uses.

Current Problems

About two thirds of the water systems reported current problems with raw water quality, seasonal water shortages, and/or water yield; a third indicated none of these was a problem at present. The incidence of particular problems increased markedly as the seriousness of urban encroachment on water supply watersheds increased, as shown in Table II-3. Where urban encroachment had become a serious problem, 56 percent of the water systems reported problems with seasonal shortages, 47 percent reported water quality problems, and 21 percent were having a problem with an insufficient yield from their watersheds.

Because a number of water systems are having yield and seasonal shortage problems, we asked them about the availability of additional sources of supply. About one out of every six water systems seeing a need for additional raw water supplies reported that no additional raw water sources were available. These water systems obviously have a very high stake in seeing that their existing sources of supply remain uncontaminated by excessive urban growth and other pollution-generating land uses. Among the water systems which reported that additional raw water sources were

35

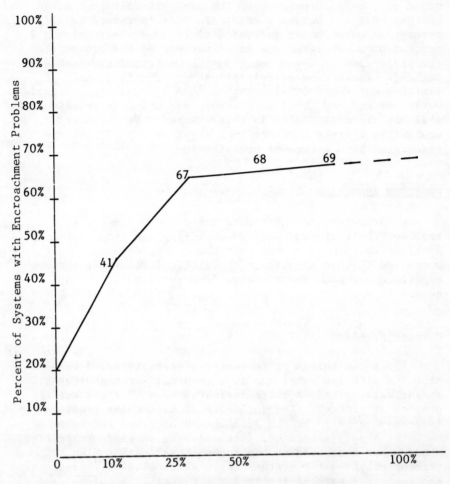

Figure II-1.

Extent of Urbanization of Water Supply
Watersheds and Perception of Urban
Encroachment as a Problem

Table II-3. Variation in Current Raw Water Problems
by Seriousness of Urban Encroachment

| | Percent of Water Systems Reporting Raw Water Problem When Urban Encroachment Reported to Be: | | | |
Raw Water Problem	Average for All Water Systems	A Serious Problem	A Problem, But Not Serious	Not a Problem
Water Quality	35	47	43	27
Seasonal Shortages	38	56	37	36
Water Yield	14	21	14	13
No Problems	35	26	34	43

available, only 15 percent felt that raw water quality would be better than current sources of supply. Fifty-five percent felt the quality of any new sources tapped would be equal to that of current sources and 30 percent—twice the proportion which saw an improvement—indicated that any new sources of raw water would be of lower quality than the raw water sources that were being used at present. Thus for a significant proportion of water systems, tapping new sources of supply will result in lower quality raw water with concomitantly increased threats to public health and potentially higher water treatment costs.*

Thirty-five percent of the national sample of water systems were already having problems with raw water quality.

---

*A number of studies have shown that water treatment costs increase as raw water quality decreases. For example, Young and his associates demonstrated with multivariate equations that turbidity, hardness, color, and biological oxygen demand are significant variables in estimating treatment costs.[3] Brandt has estimated that excessive sediment in potable water supplies increases treatment costs by $14 million (1972 dollars) per year.[4]

These problems were attributed by water system managers to both rural and urban sources of pollution:[*]

Rural Sources

47% Cultivated cropland
32% Forest activities
18% Feedlots, hog parlors, large dairy operations
7% Mining
21% Other sources

Urban Sources

28% Stormwater runoff from urban and industrial areas
26% Septic tanks
13% Municipal wastewater treatment plant discharges
8% Industrial waste discharges
6% Chemical spills

Most of these are dispersed "nonpoint" sources of pollution which cannot be readily abated through conventional wastewater collection and treatment methods.

In many rural areas, the primary source of pollutants is agricultural cropland, which may contribute sediment, nutrients, pesticides, and salts through runoff into streams and water supply impoundments. The U.S. Department of Agriculture has estimated that of the 360 to 420 million acres of cropland in the nation, 65 to 75 million acres need treatment for sediment control purposes, 100 to 125 million acres for pesticide control, 150 to 200 million acres for nutrient control, and 12 million acres for salinity control.[5] Although forests normally do not degrade water quality, logging and associated road construction can be very destructive of streams and rivers. Poor road construction on steep slopes can produce washouts, landslides, and gullying which increase stream sedimentation, and the application of pesticides may pollute nearby streams and water bodies through

---

[*]County officials and regional planners were also asked about sources polluting surface water supplies. Compared with water system managers, they tended to be more aware of urban sources of pollution (36 and 41 percent of the county officials and regional planners, respectively, named stormwater runoff as a pollution source; 51 and 37 percent named septic tanks; 25 and 35 percent cited waste treatment plant discharges; and 13 and 18 percent named industrial discharges), but their perceptions of rural pollution sources were generally similar to those of water system managers.

direct fallout during application, overland flow in solution
or attached to soil particles, and through groundwater
seepage.[6] The third most frequently cited source of rural
pollution—feedlots, hog parlors, and large dairy operations-
-stems from improper handling of the some 1.2 billion tons of
livestock wastes produced annually.[7] Finally, mining activities may contribute acids, nutrients, and a variety of metals
to the aquatic environment.

The incidence of rural pollution sources varied
significantly among federal regions. Cultivated cropland was
cited as a source of pollution by 47 percent of the water
systems nationwide, but by as many as 80 percent in Region
VII (Iowa, Kansas, Missouri, and Nebraska) and 69 percent in
Region V (Illinois, Indiana, Michigan, Minnesota, Ohio, and
Wisconsin). Similarly, although forest activities were noted
as a pollution source by 32 percent of the water systems
nationwide, in the Pacific Northwest (Region X) they were
reported by 73 percent of the water systems surveyed.
Livestock wastes were three times as likely to be mentioned
as a contributor to water quality problems in Region VII
(cited by 60 percent of the water systems) as in the
nation as a whole, and were mentioned about twice as often as
in the nation as a whole by water systems in Region V.

As urbanization increases, the significance of forest
activities as a source of pollution steadily decreases.
Surprisingly, however, other rural sources of pollution did
not vary directly with watershed urbanization. Cultivated
cropland was just as likely to be a source of pollution where
urban encroachment was a serious problem for water systems as
where it was not a problem at all. This also tended to be
true of livestock wastes and mining activities. Thus when
watersheds are undergoing urban development, it cannot be
assumed that previous rural sources of pollution automati-
cally disappear. Instead, both rural and urban pollution
sources may exist side by side, creating a double-barrelled
threat which must be addressed by watershed management
programs.

Urban water quality programs once focused almost
exclusively on municipal and industrial "point" discharges.
However, these were of much less concern to water system
managers than nonpoint urban sources, particularly stormwater
runoff and drainage from septic tanks. The problem of urban
runoff stems from two features of urban growth and
development—the removal of vegetation and earthmoving during
construction and the increase in impermeable surfaces.[8/9]
The removal of vegetation and grading during construction
increase soil erosion; in fact, on an acre for acre basis,

construction activities are the leading contributor of sediment, producing an average of 1,100 tons per acre per year.[10] Nationally, it has been estimated that one million acre feet per year of reservoir storage is lost due to sediment accumulation from urban and rural land. The estimated cost of removing this sediment is $1 billion per year.[11]

The increase in impermeable surfaces—the area of roofs, roads, parking lots, and the like—which accompanies urbanization can increase runoff by up to six times, causing an increase in erosion on unpaved portions of a site and in bank erosion in receiving streams. Stormwater running off urban surfaces is exposed to a wide variety of contaminants. These may include, in addition to soil, animal droppings, tire and vehicular exhaust residues, air pollution fallout, heavy metals, deicing compounds, pesticides, decayed vegetation, and hazardous material spills.[12] Various contaminants may be heavily concentrated in a first flush effect after the onset of rainfall so that "urban runoff can be a more serious source of water pollution than municipal sewage discharges."[13] In addition to increasing pollution, the acceleration in urban runoff results in a corresponding decrease in infiltration—the amount of water which would normally filter into the ground—which may lead to decreased streamflow during dry periods and thus decreased watershed yield.

During the early stages of watershed urbanization, septic tanks are usually the most frequently used method of handling domestic wastewater. Unfortunately, septic tanks often do not work properly. In some cases they are installed in soils which are unsuitable for on-lot sewage disposal, but even when properly installed, they tend to lose their cleansing capacity with age. As a result, a leading water quality expert has observed, "A high percentage of septic tanks fail after a short life period."[14] Replacing septic systems with sewage collection systems may lessen the problem, but sewerage systems often leak (from fractures, manholes, and sewer joints) and can induce increased density which aggravates the stormwater runoff problems discussed above.

A few water systems attributed their raw water quality problems to municipal (13 percent) and industrial (8 percent) wastewater treatment plant and untreated waste discharges. Over the past ten years, a major national effort has been devoted to abating these sources of contamination and they may become less serious over time.* The final source of

---

*On the other hand, it should be noted that a number of municipal and industrial wastewater discharges are not in

urban pollution cited, chemical spills, has drawn increased
concern in recent years as national attention has focused on
the problem of hazardous waste handling, transport, and
disposal. Commenting on this problem as it affects drinking
water supplies in Connecticut, Castorina observed,[15]

> A serious hazard has been introduced by our
> dependence on the trucking industry for overland
> movement of fuels and chemicals. Accidental spills
> are occurring at an alarmingly increasing rate....
> These spills are a major hazard because so many
> major highways either parallel or drain into the
> state's surface supplies, and it is virtually
> impossible to reroute traffic to other less
> critical areas.

Precisely because it is so difficult to deal with, the
problem of chemical spills may become increasingly serious
over time.

The incidence of urban sources of pollution was much
more evenly distributed nationwide than was the case for
rural sources. There were no significant differences among
regions in the proportion of water systems which cited
stormwater runoff, industrial waste discharges, and chemical
spills as contributors to their water quality problems.
Septic tank problems were about a third more likely than
average to be noted in Federal Regions I (New England) and II
(New York and New Jersey), while municipal waste treatment
plant discharges were about twice as likely to be cited as a
problem in the Southwest (Region VI) as in the nation as a
whole.

Table II-4 illustrates why urban encroachment is a
problem for water systems. As encroachment becomes more
serious, the proportion of water systems reporting the
existence of various pollution sources increases
dramatically. Pollution from stormwater runoff was five
times more likely to be cited as a contaminant than when
encroachment was not viewed as a problem. The reported
incidence of other pollution sources also increased—septic
tank pollution by more than two times, pollution from

------

compliance with EPA standards. For example, in February
1980, EPA estimated that 63 percent of the major municipal
treatment plants in the nation were not in compliance with
the original July 1977 statutory deadline requiring secondary
treatment or more stringent treatment necessary to meet water
quality standards.[16]

Table II-4. Variation in Urban Pollution Sources
with Urban Encroachment

| | Percent of Water Systems Reporting Pollution Source When Urban Encroachment Is: | | |
Urban Sources of Pollution	A Serious Problem	A Problem, But Not Serious	Not a Problem
Stormwater Runoff from Urban and Industrial Areas	67	39	12
Septic Tanks	40	37	15
Municipal Waste Treatment Plant Discharges	18	15	9
Industrial Waste Discharges	11	9	6
Chemical Spills	18	7	4

municipal waste treatment plants by two times, pollution from industrial waste discharges by almost two times, and chemical spills by more than four times.

Future Problems

Reports of current water supply problems are dwarfed by potential problems which water systems, county governments, and regional agencies expect over the coming ten years. While 35 percent of the water systems reported current problems with raw water quality, 55 percent expect water quality problems during the decade of the 1980s. County and regional agencies are even more pessimistic, with 72 and 89 percent, respectively, anticipating water quality problems. While 38 percent of the water systems reported seasonal raw water shortages, 56 percent expect shortages over the next decade. Again, county and regional agencies are more

pessimistic, with 63 and 86 percent expecting water supply
shortages in their jurisdictions. As with their perceptions
of current problems, water systems which were experiencing
urban encroachment on their watersheds were much more likely
than average to expect both water quality and quantity
problems by 1990.

In sum, water source protection is no longer an isolated
issue. A number of water systems are currently experiencing
raw water quality problems and water shortages. Even more
expect problems to develop in the near future. Water quality
problems arise from both rural and urban sources of
pollution. However, as a watershed develops, the urban
sources become increasingly severe, while the rural sources
do not disappear until urbanization is rather far advanced.
During the 1970s water systems and local governments across
the nation began to develop management programs to mitigate
threats to drinking water supplies. The current status of
these efforts is described in the following sections,
followed by a look at regional, state, and federal efforts to
eliminate pollution.

MANAGEMENT PRACTICES OF WATERSHED PROPERTY OWNERS

     Water supply watersheds are rarely owned in their
entirety by the water systems using them as a source of raw
water.* Instead, water systems typically own only a portion
of the watershed with the remainder controlled by numerous
land owners who may use their property for a variety of
purposes. If water sources are to be protected from
pollution, both water systems and private property owners
must take steps to eliminate potential sources of water
pollution. Actions they are taking at this time are
described next.

Management Measures Adopted by Water Systems

     Traditionally, water systems have been encouraged to
seek safe, uncontaminated sources of raw water when
developing new sources of supply and to conduct periodic
sanitary surveys to detect actual and potential sources of
pollution affecting their existing sources of supply. For
example, U.S. Environmental Protection Agency recommendations
for states, in cooperation with local water systems,

_____

     *Of the 496 water systems queried for this chapter, only
16 percent had acquired entire watersheds to protect their
raw water sources.

> Surveillance and disease prevention is recommended
> with periodic, on-site fact finding as part of a
> comprehensive sanitary survey of each public water
> supply system, from the source to the consumer's
> tap, made by a qualified person to evaluate the
> ability of the water supply system to continously
> produce an adequate supply of water of satisfactory
> sanitary quality. The US EPA suggests that the
> sanitary survey, as a minimum, cover quality and
> quantity of the source; protection of the source
> (including the watershed drainage area)....

Although a sanitary survey is one of the most basic elements
in a watershed protection program, as shown in Table II-5,
less than a third of the water systems were using sanitary
surveys to protect raw water quality.

In 1973, Robert Gerber called water system managers'
attention to the benefits of using land use measures to
protect water quality.[20] He listed four reasons for adopting
a watershed protection program:

1. The cost of treating polluted water is much
    higher than the cost of treating relatively
    pure water.

2. Increased runoff from impervious surfaces may
    lead to reduced streamflow and require higher
    costs in developing additional sources of
    supply.

3. Although polluted water may be treated, people
    are often not prepared to use it for domestic
    use.

4. Protecting water supply watersheds contributes
    to other goals, such as preserving open space,
    forests, wildlife areas, and water bodies for
    recreational use and enjoyment.

Even though water systems often do not have the authority to
enact land use controls, Gerber noted that they have the
greatest stake in seeing that a program is developed and
should play a key role in organizing and promoting watershed
protection efforts by local governments. In fact, as shown
in Table II-5, some water systems are playing the role Gerber
laid out. Thirty-seven percent of those surveyed said they
were encouraging local governments to adopt water quality

Table II-5. Measures Adopted by Water Systems to
Protect Surface Water Quality[a]

Measures	Average for All Water Systems	50,000 or More	10,000–49,999	Under 10,000
	Percent of Water Systems Adopting Measures by Size of Water System (Population Served):			
Encourage Local Governments to Adopt Measures to Protect Raw Water Quality	37	47	38	28
Acquisition of Buffer Strips Around Water Supply Reservoirs	35	46	35	25
Sanitary Surveys of Watershed	31	39	30	22
Provide Technical Assistance to Local Governments for Water Quality Protection	21	28	19	17
Acquisiton of Entire Watershed	16	13	13	23
Acquisition of Buffer Strips Along Streams Feeding Water Supply Reservoirs	13	19	13	8
None of the Above	17	14	17	19

[a]Question: What steps has your water system taken to protect
the quality of your raw surface water supplies?

protection measures and 21 percent were providing technical assistance to local governments.

In addition to conducting sanitary surveys and encouraging (and helping) local governments with land use management programs, a number of water systems have acquired strategic property to help protect water quality. As suggested above, few water systems—only 16 percent—acquired an entire watershed. On the other hand, over twice as many (35 percent) reported acquiring buffer strips around their water supply lakes and impoundments and 13 percent had acquired buffer strips along feeder streams. In general, various authorities recommend that water systems acquire a minimum of 100 feet around impoundments to buffer them from adjacent eroding land and that they acquire other lands where private owners are not providing adequate land treatment to minimize erosion and the generation of other pollutants.[21/22/23]

With the exception of acquiring entire water supply watersheds, which was most common among smaller water systems, the larger the water system the more likely it was to have employed each of the measures listed in Table II-5. Water systems' use of these measures also varied systematically among regions of the country and with the degree of watershed urbanization. In general, as urbanization proceeds, water systems are less and less likely to employ land acquisition strategies to protect water quality— probably because land becomes too expensive—and more and more likely to look to local governments for help with watershed management. For example, where urban encroachment on the watershed was viewed as a serious problem, 60 percent of the water systems reported that they were encouraging local governments to adopt water quality protection measures and 47 percent were providing local governments with technical assistance. Only half as many were encouraging local governments to adopt protective measures where urbanization was not yet a problem. Also, when water systems began having trouble with urban encroachment, they were increasingly likely to conduct sanitary surveys; however, even in these cases less than a majority reported using sanitary surveys in an effort to protect water quality.

Water systems in New England, New York, and New Jersey were much more likely than those in other regions to be taking active steps to protect raw water quality. In part, this may be due to their reliance on less than full treatment to protect public health; where water systems are not using filtration, for example, they have an added incentive to minimize watershed activities which may generate sediments

46

and other pollutants. Gerber has estimated that going from chlorination to full treatment would cost a medium-size water system with revenues of $100,000 to $1,000,000 per year an increase in production costs of 10 percent per year. For small systems, the cost increase could be as much as 25 percent, depending on the age of the system and other factors.[24]

From this brief review, it is apparent that water systems across the nation could be doing much more to protect their surface water supply sources. At present, the average system with 5,000 or more customers is devoting less than four person-weeks per year to watershed protection.* Some water systems--those in New England, those already experiencing problems from urbanization, and those with larger systems--tended to be doing more than others to guard against pollution, but even among these groups less than a majority were using any of the management measures which have been recommended by knowledgeable authorities. The most frequently employed measure, calling on local governments for help, is indirect at best and will be effective only if local governments are willing to guide growth away from water supply watersheds and regulate watershed activities to minimize pollution.

Management Measures Adopted by Private Landowners

In seeking local governmental help, many water systems look for governmental actions which will encourage (and often require) private landowners to take steps to minimize pollution from their land-disturbing activities. For agricultural activities, these steps are called "Best Management Practices" (BMPs). Best Management Practices are discussed later in this chapter with our review of the federal and state soil conservation programs which have fostered their use. The urban equivalents of agricultural Best Management Practices are erosion control and stormwater management measures designed to control the quantity and quality of stormwater running off residential, commercial, and industrial areas. As noted earlier, pollution from stormwater runoff is the number one water quality problem in urbanizing watersheds. A variety of structural works as well as various management techniques have been proposed to minimize pollution from stormwater. The frequency with which they have been adopted by landowners and developers is summarized in

_____

*In fact, a quarter of the systems studied spent only four person-days or less on watershed protection.

47

Table II-6. As can be seen, the various physical approaches (storage, infiltration, reduction in velocity of overland flow, and filtration) have been used more frequently than the municipal management measures listed at the bottom of the table.

The most commonly employed measures are those designed to reduce peak runoff and erosion by reducing the velocity of overland and channel flow. Vegetative cover and grade stabilization were often used in over half of the counties responding to our survey, while swales, diversion structures adjacent to steep slopes, and nonvegetative cover, such as mulches and straw, were used less frequently. In addition to reducing erosion, by reducing the velocity of runoff these measures are also effective in controlling other pollutants carried by stormwater runoff.

Various filtration devices, such as the use of straw or hay bales and storm drain filters during construction, are in fairly common use. These measures help reduce peak discharges and trap sediments, but often have to be cleaned after each storm event and may lose effectiveness if they are not adequately maintained. Detention basins designed to store stormwater and remove settleable materials were often used in about a third of the counties surveyed. Other storage methods, such as the use of small storage areas (earth dikes, traps, etc.), parking lot storage, and underground storage, were being used less frequently. Detention basins can be designed to reduce both peak and total discharge of stormwater and to settle out various pollutants. They require extensive maintenance, but often can serve dual purposes as stormwater management structures and recreational facilities or residential amenities (wet ponds and lakes).

Various municipal management measures were less likely to be used than the structural devices mentioned above, possibly because they require a continuing expenditure of funds and costs cannot be shifted from the public to the private sector. Frequent street cleaning to help improve the quality of stormwater runoff was often used in about a quarter of the counties surveyed, and 16 percent reported that sewer flushing during dry weather was often applied to help reduce the first flush effect of stormwater. On the other hand, sewer management for storage was infrequently used.

The private sector approaches to minimizing pollution during urban development that have been reviewed here should be used much more frequently. Because most of them increase

Table II-6. Private Sector Measures Used to Control Stormwater Quantity and Quality[a]

Stormwater Management Method	Percent of Counties Reporting Measures Used in Their Jurisdiction:		
	Often	Rarely	Never
Storage of Stormwater			
1. Detention Basin or Ponds	32	45	23
2. Small Storage Areas, Such as Earth Dikes, Traps, etc.	20	52	28
3. Parking Lot Storage	6	31	63
4. Underground Storage	4	26	70
Increase Infiltration			
1. Infiltration Trenches/Pits	10	45	45
2. Porous Pavement	3	30	67
Reduce Velocity of Overland/Channel Flow			
1. Vegetative Cover	67	24	9
2. Grade Stabilization	53	39	8
3. Swales	42	37	21
4. Diversion Structures Adjacent to Steep Slopes	25	54	21
5. Nonvegetative Cover Such as Mulch, Straw	11	59	30
Filtration			
1. Straw or Hay Bales During Construction	46	36	18
2. Storm Drain Filters During Construction	20	43	37
Management			
1. Frequent Street Cleaning	26	55	19
2. Sewer Flushing	16	63	21
3. Sewer Management for Storage	7	47	46

[a]Question: How often are the following methods used to control stormwater quantity and quality in your jurisdiction?

the costs of construction and may not add correspondingly to the value of the property being developed, there is a natural tendency for landowners to underinvest in stormwater management. For this reason, local jurisdictions across the nation are increasingly enacting stormwater management and erosion control ordinances to require landowners to minimize the adverse impacts of development on water quality. These and other possible local governmental watershed protection measures are examined next.

WATERSHED PROTECTION MEASURES ADOPTED BY LOCAL GOVERNMENTS

Local governments have been relatively slow to formulate programs to protect water resources in urbanizing areas, in part because the effects of land use on water quality are not well understood.[25] On the other hand, for over fifty years local governments have administered various measures which, whether intended or not, have affected water quality. Measures to protect public health through the regulation of septic tanks; measures to protect consumers and eliminate future costly public expenditures through regulation of the subdivision and development of property; and public acquisition of land for parks and other purposes are all familiar governmental activities which may help mitigate the adverse effects of development on water quality.* As shown in Table II-7, many such measures were being used in the counties surveyed for this study, even though only 22 percent of the counties reported that the measures they were employing had been adopted as part of a conscious watershed protection program.

The four leading sources polluting drinking water supplies were cultivated cropland, forest activities, stormwater runoff from urban and industrial areas, and septic tanks (see above, page 38). A majority of the counties had adopted measures which have the potential to mitigate each of these sources. Four of every five counties had an agricultural soil conservation and/or forest management program which was having a beneficial effect on water quality. Septic tank permits were administered by nine of every ten counties and sedimentation and erosion control regulations had been adopted by six of every ten counties. Almost as many counties, 56 percent, had adopted on-site stormwater management regulations, while subdivision regulations, which may contain provisions governing land

_____

*The history of local governments' use of land use controls is described in reference 26.

50

Table II-7. Local Governmental Measures Used to Protect
Raw Drinking Water Quality in U.S.

| | | Percent of Counties: | | |
| | | | Perceived Impact on Water Quality[a] | |
	Measure	Use Measure	Major	Minor
1.	Septic Tank Permits	90	52	48
2.	Subdivision Regulations	83	31	69
3.	Agricultural Soil Conservation and/or Forest Management Program	80	34	66
4.	Zoning Regulations	63	35	65
5.	Sedimentation and Erosion Control Regulations	60	28	72
6.	Treatment Standards for Point Sources of Pollution Within Water Supply Watersheds	57	49	51
7.	On-site Stormwater Management Regulations	56	27	73
8.	Preferential Taxation to Maintain Agricultural Uses or Open Space	41	17	83
9.	Locating New Roads, Water, Sewer Lines out of Watershed to Discourage Development	40	27	73
10.	Critical Areas Designation and Regulation			
11.	Fee Simple Acquisition of Watershed Land	27	59	41
12.	Acquisition of Easements or Development Rights on Watershed	23	26	74
13.	Special Water Supply Watershed Ordinances	19	42	58

[a]Percent of counties using measure.

disturbing activities, had been adopted by 83 percent of the counties. Fewer counties, though still a majority (63 percent), had adopted zoning regulations, which can be used to keep potentially polluting activities away from watercourses and to restrict activities involving hazardous material from locating in water supply watersheds. Other methods of protecting watersheds--such as preferential taxation to preserve agricultural land uses, capital improvements programming of roads and water and sewer lines outside of the watershed to induce development to locate elsewhere, and land acquisition--had been adopted by less than a majority of the counties responding to our survey. However, land acquisition was one of only two measures--the other was septic tank regulations--judged by a majority of the respondents to be having a "major" impact on water quality. Most of the measures, as shown in Table II-7, were viewed as having a minor effect.

Obstacles to Watershed Management

The relative ineffectiveness of many land use measures may be due to a variety of factors. In most cases measures had not been adopted as part of a conscious watershed protection program. Thus, any positive effects on water quality may have been fortuitous rather than planned. Septic tank regulations, for example, have often been designed and administered to prevent nuisance situations from arising and to protect public health, not to protect water quality.[27] Unless septic tank regulations are designed to keep on-site disposal systems from locating on poor soils, steep slopes, and near watercourses, all of which are likely to impair their proper operation, and to monitor their proper functioning (rather than merely respond to nuisance complaints), they may not be very effective in protecting water quality. A number of other factors which local officials identified as obstacles to water supply watershed management are summarized in Table II-8.[*]

---

*Other authors have identified obstacles to environmental management programs which are also instructive. For example, according to Jon Kusler, "Major impediments to successful efforts include: (1) inadequate funding or a defeatist attitude held by the program staff; (2) inadequate leadership; (3) failure to involve local units of government; (4) inexperienced and unimaginative program staff; (5) uncooperative and inflexible attitudes in state program administration...; (6) poor public education and public

Table II-8. Local Officials' Assessments of Obstacles to Expanded Water Supply Watershed Management Program[a]

Obstacles	Percent of Local Jurisdications Assessing Obstacles As:		
	Serious	Moderate	Minimum
Political Obstacles			
Opposition by Major Industries	73	21	6
Opposition by Agricultural Interests	56	32	12
Opposition by Building, Real Estate, Land Development Interests	42	32	26
Lack of Interest by Elected Officials	38	44	18
Watershed Protection Not Perceived as a Problem	31	47	23
Administrative Obstacles			
Legal Constraints in State Law	65	22	13
Watersheds Are Too Large	47	33	19
Watersheds Extend Over More Than One Local Governmental Jurisdiction	33	36	31
Lack of Financial Resources to Undertake/Expand Watershed Protection Program	15	31	54
Technical Obstacles			
Insufficient Technical Assistance Available from Federal, State, and Regional Agencies	50	37	13
Lack of Appropriate Professional Personnel	45	38	17
Lack of Needed Technical Information	44	38	18

[a]Question: How serious are the following obstacles to expanded governmental action in water supply watershed management?

The most serious perceived obstacles are political and administrative in nature. Opposition by major industries was cited most often as a serious problem, and a majority of officials also felt that opposition from agricultural interests had limited the development of watershed protection programs. The most serious administrative obstacles were legal constraints in state laws and difficulties encountered in attempting to manage very large watersheds. In addition to political and administrative problems, half of the officials cited insufficient technical assistance from state and federal agencies as a serious obstacle, and over 40 percent cited the lack of appropriate professional personnel and needed technical information as serious problems.

Earlier we noted that when water systems perceived watershed urbanization as a problem, they were increasingly likely to begin conducting sanitary surveys and to encourage local governments to take action to protect water quality. On the other hand, they were much less likely to be employing land acquisition strategies to protect their drinking water sources. Local governments acted in a similar fashion. The proportion of counties adopting land use regulations increased sharply as urbanization became a problem:

Land Use Regulations	Percent of Jurisdictions Adopting Measure by Status of Urbanization	
	Serious Problem	Not a Problem
Subdivision Regulations	84	59
Zoning Regulations	78	52
On-site Stormwater Management Regulations	54	37

---

involvement; (7) enabling statutes with weak definition criteria and value outlines of agency or local government responsibilities; (8) expenditures of scarce funds on small scale maps and other data of little use in program implementation; (9) inability or unwillingness to carefully evaluate development permits; (10) failure to promulgate rules and guidelines for the processing of individual development permits; (11) political pressures which result in policies or development permits inconsistent with natural resource values and hazards."[28]

On the other hand, other measures were less likely to be used. For example, as was the case with water systems, local governments were less and less likely to be using land acquisition to protect water quality as urbanization advanced. Critical areas programs and special water supply watershed ordinances were also used less frequently after urbanization had become a problem:

Watershed Protection Measure	Percent of Jurisdictions Adopting Measure by Status of Urbanization	
	Serious Problem	Not a Problem
Land Acquisition	35	45
Special Watershed Ordinance	25	40
Critical Areas Regulation	29	34

It appears likely that when urbanization gains momentum, the potential for profit from land value appreciation and continued growth becomes so great that programs with the potential to limit growth to protect water quality become increasingly difficult to enact. On the other hand, regulations which are designed to protect water quality by affecting the way land is developed, but which may allow continued growth and development of the watershed, are well accepted, as indicated by the increasing use of measures such as subdivision regulations and on-site stormwater management regulations. Zoning regulations--particularly large lot zoning and prohibitions of particularly hazardous land uses--can be used to limit urbanization of the watershed. However, zoning regulations are often relaxed as development pressures mount, which may account for the greater acceptance of zoning compared with special watershed ordinances and critical areas regulations. These latter measures are usually designed specifically for resource protection and as a result are less likely to be modified to accommodate urban development.

In terms of lessons for watershed management, these findings suggest that local governments need to become concerned about protecting water supplies before urbanization becomes a problem (before even 10 percent of a watershed is developed for urban use). If local governments delay action, as shown here the use of several potentially effective management tools may be foreclosed.

Before concluding this discussion, it should be noted that the use of measures which are often mandated in state legislation--septic tank regulations, sedimentation and erosion control regulations, and agricultural soil conservation and forest management programs--did not vary with the level of watershed urbanization. Each of these measures was in place in a majority of the counties which were not experiencing problems with urbanization and also in a majority of those where urbanization was a serious problem. This suggests that state regulation or state-mandated local regulation may be an effective means of protecting water supply watersheds and insulating the watershed protection program to some degree from attack as urban development pressures mount.

Overall Program Effectiveness

In addition to asking about the use and effectiveness of individual watershed management measures, such as zoning ordinances and septic tank permits, we also asked both water system managers and local officials to assess the overall effectiveness of the entire array of measures that were being used by local governments to protect drinking water quality. As illustrated in Table II-9, both sets of respondents tended to view water supply watershed management as moderately or slightly effective. Only one in four of the water system managers and one in eight of the local government officials rated the local watershed management effort as very effective. About one in five of each group thought their local program was ineffective.

Overall effectiveness was examined in relation to the types of drinking water sources needing protection, sources of pollution existing in the watersheds, types of measures which had been adopted to protect drinking water quality, and obstacles which had been encountered. In the first case, no association was found between types of drinking water sources and program effectiveness. Thus, local jurisdictions relying on streams as their source of supply were no more nor less successful in protecting their supply from contamination than those drawing raw water directly from lakes or those using groundwater.

Watershed management programs tended to be less effective when they had to contend with certain sources of pollution. For example, when major point sources of pollution--municipal waste treatment plants, industrial discharges, and feedlots, hog parlors, and large dairies-- were contributing to water quality problems, programs tended

Table II-9. Overall Evaluations of Local Government
Watershed Protection Efforts[a]

Overall Evaluation of Local Government Watershed Protection Measures	Water System Managers	Local Government Officials
Very Effective	24	13
Moderately Effective	30	41
Slightly Effective	25	27
Ineffective/No Measure Used to Protect Raw Water Quality	21	19

[a]Question: How would you rate the overall effectiveness of
the measures local governments have used to protect drinking
water quality?

to be judged as less effective. Local jurisdictions where
cultivated cropland was a major source of water quality
problems also tended to view their programs as less
effective. On the other hand, the existence of most other
nonpoint sources of pollution—septic tanks, stormwater
runoff from urban and industrial areas, forest activities,
and chemical spills—was not associated, either positively or
negatively, with program effectiveness ratings. In terms of
lessons for future watershed management, these findings
suggest that if at all possible major point sources of
pollution, such as large industrial plants, should be
discouraged from locating in water supply watersheds. In
addition, however, the data show that some rural uses,
particularly cultivated cropland and livestock operations,
may be as difficult to deal with in water quality protection
programs as urban development.

Table II-10 shows how evaluations of overall program
effectiveness improved with the adoption and individual
effectiveness of various management measures. In most cases,
the proportion of managers rating the overall management
program as very effective did not change markedly when
measures were adopted, but were judged to have only a minor
impact on water quality. Thus, as was noted above of septic

57

Table II-10. Improvement in Overall Program Evaluations
with Adoption and Effectiveness of Individual
Management Measures

| Local Government Watershed Protection Measures | Percent of Water System Managers Rating Overall Local Government Management Program as Very Effective by Use and Impact of Measures: | | |
| | Measure Not Used | Measure Used and Impact Is | |
		Minor	Major
Septic Tank Permits	25	29	46
Subdivision Regulations	19	19	37
Agricultural Soil Conservation/ Forest Management Program	14	22	36
Zoning Regulations	19	18	40
Sedimentation and Erosion Control Ordinance	17	24	33
On-site Stormwater Management	20	14	46
Locating New Roads, Water, Sewer Lines Out of the Watershed	19	24	39
Critical Areas Designation/ Regulation	17	23	37
Land Acquisition	17	23	37
Special Water Supply Watershed Ordinance	16	18	44

tank regulations, the mere existence of regulations without
careful attention to how they are designed and used to
protect water quality usually will not lead to a better
watershed management program. On the other hand, the data in
Table II-10 illustrate that when measures are carefully
formulated to protect water quality, they can lead to a much

more effective overall management program. In most cases the
proportion of water system managers rating local governments'
watershed protection efforts as very effective overall more
than doubled when each measure was adopted and was felt to
have a major impact on water quality.

Earlier we noted that only one in five (22 percent) of
the local governments surveyed had adopted watershed protection measures as part of a coherently designed program established specifically to protect water supplies. However, this
step has a strong impact on the overall effectiveness of the
watershed protection measures that are adopted. Table II-9
showed that only 13 percent of the local government officials
surveyed rated watershed protection measures as very effective overall in protecting water quality. This proportion
increased two and one half times to 32 percent of the
officials from jurisdictions where measures had been adopted
as part of an identifiable watershed protection program.
Thus, just as individual measures have more effect when they
are designed and adopted to protect water quality, the overall array of measures is more effective when it is put
together specifically to protect raw drinking water quality.

The contribution of individual measures to overall
program effectiveness was gauged using the statistic Gamma,
an ordinal measure of association. Based on the Gamma
statistic as an indicator of their contributions to overall
program effectiveness, the most important individual
management measures are:

Most Important Management Measures (in order)

1. Land acquisition

2. Special water supply watershed ordinances

3. Critical areas designation/regulation

4. Subdivision regulations

5. Locating new roads, sewer lines, water lines
    outside of the watershed

6. Agricultural soil conservation/forest
    management programs

7. Sedimentation and erosion control ordinance

8. Zoning ordinance

9. On-site stormwater management

10. Septic tank permits

Since land acquisition, special water supply watershed ordinances, and critical areas designation and regulations were among the least used measures (see Table II-7), their adoption by a larger number of local governments is a straightforward means of improving local watershed management.

In addition to examining the watershed management measures most associated with effective protection of water quality, we looked at how the presence of various obstacles affected overall ratings, again using the Gamma statistic as a measure of association. The obstacles most likely to lead to an ineffective set of watershed protection measures are:

Most Important Obstacles to Program Effectiveness
   (in order)

1. Watershed protection not perceived as a problem

2. Lack of interest by elected officials

3. Lack of appropriate professional personnel

4. Opposition by agricultural interests

5. Lack of needed technical information

6. Lack of financial resources

7. Insufficient technical assistance from federal, state, and regional agencies

8. Opposition by building, real estate, and land development interests

9. Watersheds extend over more than one local jurisdiction

10. Legal constraints in state law

These obstacles can be viewed as a checklist of factors which need to be considered in devising programs to protect water supply sources from contamination. First and foremost they suggest that programs will achieve little success, and probably not even get off the ground, unless people are convinced a problem exists and political support can be marshalled. In addition to gaining support from local elected officials, the data also suggest that potential opposition from groups which stand to gain from watershed development needs to be prevented from forming and counteracted if it does arise. These groups include not only those directly associated with urban development, such as land developers, but also farmers and other interests which benefit from land value appreciation in urbanizing areas.

Administrative problems had less impact on program effectiveness than might have been supposed from the large proportion of officials who cited them as a problem (see Table II-8). Nevertheless, one in particular should be reemphasized--intergovernmental relations. As shown above, watersheds extending over more than one local jurisdiction ranked ninth among the top ten obstacles to an effective management program. Almost 80 percent of the water systems surveyed had watersheds which straddled local governmental jurisdictions. Only a third of the county officials surveyed said they did not have a problem with intergovernmental relations. Thus, in mustering political support for watershed protection and management, water systems must pay close attention to building cooperative relationships among all of the local governmental entities involved.

The need for additional professional personnel, additional technical information and help from state and federal agencies was mentioned as a serious obstacle to program effectiveness by between 40 and 50 percent of the counties surveyed (Table II-8). As shown above, the need for professional personnel and for more technical information ranked among the top five factors in their statistical association with ineffective programs. Local officials were asked what types of technical information would enable them to do a better job with watershed management. Their responses and the proportion of officials mentioning each type of information were:

Type of Information	Percent of Officials
1. Better information on land use-water quality relationships*	58

---

*The lack of adequate information on land use-water quality relationships is also noted from time to time in the literature. According to Coughlin and Hammer, for example, one of the reasons the attempt to develop a plan for protecting the Brandywine stream corridor in Pennsylvania was

61

2. Information on management alternatives
   for addressing specific problems   51

3. Hydrologic data   42

4. Water quality data   40

5. Land use data   16

6. Soil survey data   16

7. Topographic data   8

Only 15 percent of these officials said that no additional
technical information was needed to make their watershed
management efforts more effective.

The provision of technical information and data is one
of the major functions of regional, state, and federal water
quality programs. The fact that half of the local officials
queried indicated that the lack of such data was a major
problem (Table II-8) and that those voicing this concern
tended to have ineffective programs suggest that these
federal and state agencies could be doing a much better job
of providing technical support for local agencies. In
addition, local governments' list of information needs indicates that more than scientific data is needed. Information
about appropriate management alternatives ranked second on
their list. Agencies providing technical support to local
governments should take note of this need.

To summarize briefly, we have seen in this section that
local governments' watershed protection efforts have yet to
achieve high effectiveness ratings from either local government officials or water system managers. However, a number
of factors which account for the low ratings have been
identified. They include: (1) the fact that watershed
management measures were often adopted for reasons other than
water quality protection, so that in most cases they have
only a minor impact on water quality; (2) the failure of
water systems and local governments to formulate watershed

---

a failure was inadequate scientific information linking
improvements in stream quality with the land use measures
proposed in the Brandywine Plan.[29] Similarly, the U.S.
General Accounting Office has pointed out the lack of cause
and effect data about various pollution control measures,
noting that this may hinder implementation and lead to legal
challenges.[30]

protection programs designed specifically to protect water quality; (3) the infrequent use of measures--such as special watershed ordinances--with the greatest potential for contributing to an effective management program; and (4) the existence of political, administrative, and technical obstacles to program effectiveness. None of these problems is insoluble. In fact, most can be avoided once local governments decide that protecting water supply sources from contamination is an important goal and begin channelling local resources to achieve this end. In addition, the methodology and other information provided in this Guidebook should help to make their efforts more efficient and, hopefully, more effective as well.

The Payoffs from an Effective Watershed Management Program

The formulation of an effective watershed protection program can produce important benefits for water systems and the people they serve. Those already being realized and likely to be realized in the future by the water systems participating in our survey are indicated in Table II-11. The table summarizes the current and expected incidence of various raw water problems where local governments have developed very effective, moderately effective, and ineffective overall watershed protection efforts.

In the case of current problems, it can readily be seen that the proportion of water systems reporting no raw water problems almost doubled--from 26 percent to 48 percent--when program effectiveness ratings improved from "ineffective" to "very effective." The greatest impact of watershed protection programs is on raw water quality. Where watershed protection efforts were rated as ineffective, 54 percent of the water systems reported water quality problems; where a very effective watershed protection program was in place, only 19 percent reported raw water quality problems. On the other hand, watershed protection programs had no apparent effect on the yield obtained from watersheds and only a slight (and statistically insignificant) effect on current seasonal shortages.

Development of an effective watershed management program also had little effect on water system managers' perceptions of future raw water shortages. About the same proportion--55 to 57 percent--expected shortages to occur regardless of whether their local governments had developed an effective or ineffective watershed management program. However, a number of other future benefits were evident. For one, water system managers were much less likely--42 percent vs. 71 percent--to

Table II-11. Program Effectiveness and Raw Water Problems

Current and Future Raw Drinking Water Problems	Percent of Water Systems Experiencing or Expecting Raw Water Problems by Evaluation of Program Effectiveness		
	Very Effective	Moderately Effective	Ineffective
Current Problems			
Water Quality	19	39	54
Water Yield	10	16	11
Seasonal Shortages	39	35	46
None of Above--No Problems	48	36	26
Future Problems (by 1990)			
Deteriorating Quality of Surface Water Used for Raw Water Supply	40	61	72
Lack of Acceptable Alternative Sources of Raw Water	42	61	71
Loss of Reservoir Storage Due to Sediment	27	41	56
Chemical Spills	16	33	38
Raw Water Shortages	55	57	56

feel that finding an acceptable alternative source of supply was going to be a problem if a very effective management program was in place rather than an ineffective program. As illustrated in Table II-11, they were also much less likely to expect problems by 1990 with deteriorating raw water quality, loss of reservoir storage from sediment, and chemical spills.

In short, water systems now and in the future expect local government watershed management programs to pay off. These payoffs include reduced threats to public health from using contaminated raw water, decreased costs of treating pure rather than polluted water, and less likelihood of incurring the added costs of going far afield to find an acceptable future source of supply when the safe yield of the current source is exceeded.

REGIONAL AGENCIES AND WATERSHED PROTECTION

In our federal system, most public functions tend to be shared by all levels of government. In the case of drinking water source protection, the major management responsibilities lie with water systems and local governments, but regional agencies, state governments, and the federal government are also involved in our national effort to provide safe drinking water to the public. Regional agencies were drawn into water quality protection by Section 208 of the Federal Water Pollution Control Act Amendments of 1972. Section 208 established a mechanism for funding water quality planning programs in each state and over 200 substate regions across the United States.

In August 1975, the U.S. Environmental Protection Agency issued guidelines to 208 agencies on "Water Supply and Ground Water Considerations in Preparation of a 208 Wastewater Management Plan." The guidelines directed the 208 planning agencies to: (1) develop an inventory of water systems serving or taking water from the study area; (2) project future water demands; (3) collect data on potential water supply sources to meet future demands, and determine to what extent 208 planning projects would affect existing and potential water supply quality and quantity; (4) determine which water supply agencies were to be involved in the planning projects; and (5) evaluate the effects of 208 planning projects on groundwater. These guidelines provide a clear mandate to consider drinking water supply protection in 208 water quality planning, but as Goldrosen has observed, they provided "...little guidance on how to balance that issue with other elements of the 208 program."[31]

The measures regional agencies were pursuing to protect water supply sources are summarized in Table II-12. Only two measures—providing technical assistance to local governments and including watershed protection in regional plans—had been adopted by a majority of the 549 regional agencies responding to our survey. In addition, a sizable proportion of these agencies—40 percent—were using the A-95 review

Table II-12. Measures Used by Regional Councils of
Government to Protect Water Supply Watersheds

| | Percent of Regional Agencies: | |
Measures	Use Measure[a]	Rate Measure as Most Effective in Preventing Quality Degradation
Provide Technical Assistance to Local Governments	76	48
Include Watershed Protection in Regional Plans	54	5
Discourage Public Investment from Locating in Watersheds through A-95 Review	40	8
Provide Public Information about Water Supply Watershed Protection Methods	38	5
Conduct Sanitary Surveys to Detect Pollution Sources	25	1
Monitor and Evaluate Local Watershed Protection Programs	25	4
Coordinate Local Watershed Protection Programs	24	4
Regulate Watersheds to Prevent Pollution	17	8
Designate Watersheds for Future Water Supply Use	16	2
Acquire Water Supply Watershed Land for Open Space/Park/Recreation Use	12	2
Conduct Watershed Surveys to Monitor Compliance with Existing Regulations	7	7
No Measures in Use	11	NA

[a]Question: Which of the following methods are used by your agency to help protect water supply watersheds?

[b]Question: From the items circled (methods used), select the number of the method which you feel has been most effective in preventing the degradation of drinking water supplies?

process to discourage various public investments in water supply watersheds and 38 percent were providing public information about watershed protection methods. Although many local agencies noted the existence of intergovernmental problems in pursuing watershed management programs, suggesting that intergovernmental coordination could be an important regional function, only 24 percent of the regional agencies were filling this role. In addition, about a quarter were conducting sanitary surveys to detect pollution sources in water supply watersheds and a quarter were monitoring local watershed protection efforts. Other potential regional activities--including regulating watershed development to prevent pollution, designating watersheds for future water supply use, acquiring property, and monitoring compliance with regulations--were being pursued by less than a quarter of the regional agencies.

Although 85 percent of the local governments contacted indicated that more technical information would help them improve their watershed management efforts, providing technical assistance to local governments was felt by almost half of the regional agencies to be their most effective program for preventing water quality degradation. After this, as shown in Table II-12, each of the other regional programs we asked about drew only a few votes as the most effective water quality protection measure.

When regional agencies were not constrained to choose from among the programs they were actually pursuing but instead were asked to indicate what would be the most effective measures they could adopt, regardless of whether they were currently using them or not, a completely different ordering appeared. The top five regional agency nominations for measures with the most potential for preventing the degradation of drinking water supplies were, in order:

Regional Measures with the Most Potential for Protecting Water Supply Sources	Percent of Nominations
1. Regulating existing water supply watersheds to control pollution	27
2. Regulating future/potential water supply watersheds to control pollution	18
3. Acquiring water supply watershed land for open space/park and recreational use	11

67

4. Providing technical assistance to local
   governments   9

5. Conducting watershed surveys to monitor
   compliance with existing regulations   8

Of course, few regional councils of government have been
delegated land use regulatory power by state governments, and
the prospect for this occurring in the near future is not
particularly bright. Since regional agencies cited lack of
financial resources as the number one obstacle to expanding
their watershed protection efforts (mentioned by 68 percent
as a serious obstacle), funding of regional land acquisition
also does not appear likely in many cases. In short, the
programs regional agencies feel would be most effective in
protecting water quality are for the most part beyond their
reach. This leaves technical assistance as the one measure
many agencies are currently pursuing with what they believe
to be some success and which ranks among the top five
regional programs in terms of its potential effectiveness in
preventing water quality degradation. As discussed above,
increased attention to technical assistance would also likely
draw considerable local government support.

STATE AND FEDERAL PROGRAMS TO PROTECT WATERSHEDS

Most states and the federal government do not have
programs for watershed protection per se, but many of their
efforts to ensure safe public drinking water and to protect
water quality and prevent soil erosion can contribute to
effective local watershed management. In this section we
will review briefly the major types of state and federal
programs which have been enacted and local and regional
agencies' assessments of their role in local watershed
protection.

State Programs

One of the most important roles of state government is
the provision of legislation enabling water systems and local
governments to take action to minimize threats to water
quality. Most states have enacted enabling legislation
authorizing the adoption of various land use management
measures. For example, according to Kusler forty-eight
states authorize cities and forty authorize counties to adopt
zoning controls; forty-one states authorize cities and thirty
authorize counties to adopt subdivision regulations.[32] In
addition, a number of states have authorized localities,

including water systems, to enact various critical areas
regulations, such as special watershed ordinances, and some
have given cities and, less frequently counties, home rule
powers which allow jurisdictions to adopt critical areas
programs in the absence of specific state enabling
legislation. In spite of the generally broad land use
management powers granted to local governments by most
states, as we saw above, almost two thirds of the local
governments we queried cited legal constraints in state law
as a major obstacle to effective watershed management.
Obviously, there seems to be a need for states to examine
their enabling legislation to be certain that arbitrary curbs
on local governments' authority to protect water quality are
removed.

Beyond the general issue of state enabling legislation,
a number of states have enacted specific programs which can
help localities with watershed management. These programs
and local and regional agencies' assessments of their impacts
are summarized in Table II-13. The most frequently cited
state program and the program most often viewed by local
governments as making a major contribution to effective
watershed management was state standards for septic tanks.
Another wastewater management program, state point source
permits (under the National Pollution Discharge Elimination
System established by the Water Pollution Control Act Amendments of 1972 (and 1977)) was cited by over 90 percent of the
regional agencies surveyed, of which more than a third felt
the program was making a major contribution to watershed
management. Local agencies, as shown in Table II-13, were
less likely to be aware of the point source permitting
program or rate it as highly.

A high proportion of regional and local officials were
aware of state agricultural extension and forest extension
technical services. These services are designed to inform
farmers and forest owners of Best Management Practices
applicable to their property. In both cases they may include
practices designed to minimize soil erosion and also water
pollution from fertilizers, pesticides, and animal wastes.
However, as indicated by Table II-13, although these programs
are widely available, most local and regional officials do
not believe they have made a major contribution to effective
watershed management.

A number of pesticides have been identified as
potentially carcinogenic. They may pollute drinking water by
running off farm and forest lands immediately after
application or they may settle in the soil and run off at a
later point in time when the soil is disturbed. A number of

Table II-13. State Government Programs Contributing to Effective Water Supply Watershed Management[a]

| | Percent of Local and Regional Agencies Reporting: | | | |
| | Program Applied to Juris- | Contribution to Effective Management | | |
State Programs	diction	Major	Moderate	Minimal
Septic Tank Standards				
Local Perceptions[b]	92	34	47	19
Regional Perceptions[c]	91	22	47	31
Agricultural Extension Technical Services				
Local Perceptions	85	12	54	34
Regional Perceptions	90	19	47	34
Pesticide Regulations				
Local Perceptions	81	9	51	40
Regional Perceptions	83	15	37	48
Point Source Permits				
Local Perceptions	79	28	38	34
Regional Perceptions	93	37	24	39
Erosion and Sedimentation Control Regulations				
Local Perceptions	78	17	42	41
Regional Perceptions	78	44	40	16
Forest Extension Technical Services				
Local Perceptions	71	4	42	54
Regional Perceptions	75	8	41	51
Critical Areas Regulations				
Local Perceptions	55	33	58	9
Regional Perceptions	53	9	36	55

[a]Question: Have state programs which have been applied in your jurisidiction made major, moderate, or minimal contributions to the effective management of water supply watersheds?
[b]Responses of county government officials.
[c]Responses of regional councils of government officials.

states have legislation regulating pesticides, but most states' laws govern the proper labeling and sale of these substances rather than establish rules regarding their application and effect on the environment.[33] Four out of five local and regional officials queried felt that state pesticide laws applied to their jurisdictions, but most did not feel these laws were contributing much to watershed management.

In response to the growing recognition that nonpoint sources of pollution are a serious problem in urban areas, a number of state governments have enacted erosion and sedimentation control legislation. These programs are usually one of two types: state enabling legislation for locally run programs and direct state regulation of land disturbing activities. As shown in Table II-13, almost 80 percent of both the local and regional officials responding to our surveys felt that state erosion and sedimentation control programs applied to their jurisdictions, but they were sharply divided regarding the contribution of the state programs to effective watershed management. Regional agencies rated these progams as the leading state program in this regard, while local officials were much less sanguine about these programs' effects. Since the programs are usually administered at the local level, we might assume that local officials' perceptions would be most accurate.

The final type of state program we asked about—critical areas programs—also resulted in opposite perceptions by local and regional officials, only in this case local officials were those most likely to feel that the programs were making a major contribution to watershed management. Since these programs are often administered at the regional level (although usually not by regional councils of government), it seems possible that regional officials' perceptions are most accurate here, although this would lead again to discounting the effectiveness of the programs in helping to protect water supply watersheds.

Federal Programs

Local and regional officials' perceptions of federal programs are summarized in Table II-14. A majority of the local and regional officials queried were aware of each of the federal programs they were asked about (see Table II-14), but in most cases they did not view the programs as making a major contribution to watershed management. The major exceptions were programs operated by the U.S. Soil Conservation Service (SCS). Established by the Soil

Table II-14. Federal Government Programs Contributing to
Effective Water Supply Watershed Management[a]

Federal Programs	Percent of Local and Regional Agencies Reporting:			
	Program Applied to Juris-diction	Contribution to Effective Management		
		Major	Moderate	Minimal
Soil Conservation Service Programs				
Local Perceptions[b]	96	30	50	20
Regional Perceptions[c]	98	40	46	14
Agricultural Stabilization and Conservation Service Programs				
Local Perceptions	85	21	51	28
Regional Perceptions	88	28	47	25
Environmental Protection Agency Programs				
Local Perceptions	82	17	37	46
Regional Perceptions	92	34	40	26
HUD 701 Planning Assistance Programs				
Local Perceptions	66	5	29	66
Regional Perceptions	NA	NA	NA	NA
Farmers Home Administration Programs				
Local Perceptions	NA	NA	NA	NA
Regional Perceptions	81	14	36	50
U.S. Geological Survey Programs				
Local Perceptions	NA	NA	NA	NA
Regional Perceptions	82	13	39	48

[a]Question: Have federal agencies operating within your
jurisdiction made major, moderate, or minimal contributions
to the effective management of water supply watersheds?

[b]Responses of county government officials.

[c]Responses of regional councils of government officials.
NA=Not ascertained.

Conservation Act in 1935, this program nearly blankets the nation, covering 98 percent of privately held land and providing assistance in nearly 3,000 counties.[34]

The major mechanism through which the SCS operates is soil conservation districts. These are established by state governments to help focus attention on land, water and related resource problems, to develop programs to help solve these problems, and to coordinate public and private conservation efforts. Districts provide technical assistance to land users and aid them in developing a conservation plan and installing planned conservation practices. The SCS and related soil conservation districts have been at the forefront of efforts to mitigate rural and urban nonpoint sources of water pollution through the promotion of wider use of "Best Management Practices" (BMPs). Best Management Practices have been defined for a variety of types of rural and agricultural land. They range from conservation cropping systems, contour farming and similar measures applicable to cropland; the development of ponds, planting schemes and the like for pastureland; range seeding and grazing systems for grazing land, reclamation plans for mine areas, and woodland improvement, tree planting, access road standards and other measures for woodlands. In addition to BMPs for rural areas, the SCS has developed measures applicable to already developed and developing urban areas. In fact, in a number of states soil conservation districts have assumed responsibility for administering state-mandated erosion and sedimentation control regulation of new urban development. As shown in Table II-14, SCS programs are those most recognized by local and regional officials and those most likely to be judged to be making more than a minor contribution to watershed management program effectiveness.

Because watershed management is likely to be a shared responsibility of the water systems using the watershed as a source of supply, local governments with jurisdiction over the watershed, and regional, state, and federal agencies with programs designed, at least in part, to help mitigate water pollution, we concluded the questionnaires directed to local governments and regional agencies by asking for an overall appraisal of how well this shared watershed management system was working in actual practice. Their perceptions were quite similar. As shown in Table II-15, neither group felt watershed management was very effective, nor were they likely to rate it as wholly ineffective. Instead, both local officials and regional officials tend to feel that the system is moderately effective (about half of each group) or slightly effective (about a third of each group).

73

Table II-15. Evaluation of the Combined Effectiveness of Agencies Contributing to Water Supply Watershed Management

| | Percent of Officials | |
Rating	Local Governments[a]	Regional Councils of Government[b]
Very Effective	10	9
Moderately Effective	53	46
Slightly Effective	30	34
Ineffective	7	10

[a]Question: How would you rate the combined effectiveness of the agencies involved in the management of water supply watersheds within your jurisdiction in protecting raw drinking water quality?

[b]Question: How would you rate the overall management of water supply watersheds in your region in preventing the degradation in raw water quality?

SUMMARY

In this chapter we have examined in some depth the current status of watershed protection as it is being practiced by a variety of public agencies and private landowners. While it is clear that we are still a long way from effective management of water supply watersheds in most areas of the country, a start in this direction has been made. A number of water systems are aware that they have problems with raw water quality and that even more serious problems are on the horizon. Unfortunately, fewer water systems have taken action to protect raw water quality. Water systems tend to become aware of the need to take positive action only after they realize watershed urbanization is a problem, but at this point some of the most effective steps they might take, such as land acquisition and the establishment of special watershed protection ordinances, are increasingly difficult to adopt. As urban-related problems mount, water systems are increasingly likely to turn to local governments for help.

Local governments are operating a variety of measures which have some effect in mitigating potential sources polluting raw drinking water supplies. In most cases these measures were not adopted as part of a carefully thought out and designed watershed protection program. Because of this ad hoc approach, few local government efforts were judged as very effective in protecting water sources from contamination. This and other factors detracting from the effectiveness of watershed management programs were identified in this chapter. They constitute a check-list of considerations which must be addressed in the future as new programs are formulated and steps are taken to improve the performance of existing programs.

Taking the effort to make watershed management effective has a number of benefits for water systems and the people they serve. In this chapter we have seen how the incidence of a broad array of water problems--from poor water quality to the unavailability of acceptable future sources of supply--decreases with the development of a very effective management program.

Finally, it should be clear that protecting sources of drinking water from contamination is a shared responsibility. Because they have the most direct stake in watershed management, water systems must take the lead in adopting watershed protection measures which are within their powers and capabilities and encouraging others to take appropriate actions as part of a coordinated management system. In addition to examining how local governments have helped with watershed management, in this chapter we have also identified potentially useful roles for regional agencies--particularly in providing technical assistance to water systems and local governments--and we have seen that a variety of state and federal programs can also make a contribution.

This chapter has reviewed and drawn lessons from how water supply watershed management has been practiced in the past. From this review we can only conclude that improvements can be made, and that they are worth pursuing. In the next five chapters of this Guidebook, we describe a systematic approach to formulating a watershed management program, from isolating the critical water quality problems, the subject of the following chapter, to evaluating and revising on-going programs, which is the concluding step in our methodology described in Chapter VII.

REFERENCES

1. Dillman, Don A. Mail and Telephone Surveys: The Total
    Design Method, John Wiley and Sons, Inc., New York,
    1978.

2. Bureau of the Census, U.S. Department of Commerce.
    Statistical Abstract of the United States: 1981, U.S.
    Government Printing Office, Washington, D.C., December
    1981.

3. Young, G. Kenneth, Jr., Theodore Popochak, and George H.
    Burke, Jr. "Correlations of Degree of Pollution with
    Chemical Costs," Journal of the American Water Works
    Association, Vol. 57 (1965), pp. 293-297.

4. Brandt, G.H. et al. An Economic Analysis of Erosion and
    Sediment Control Methods for Watersheds Undergoing
    Urbanization, Dow Chemical Corporation, Midland, Mich.,
    1972.

5. U.S. Department of Agriculture, Soil Conservation
    Service. Environmental Impact Statement, "Rural Clean
    Water Program" (draft), June 16, 1978, p. 16.

6. Meier, Marvin C. "Watershed Management--Or Regulation,"
    Journal of the Utah ASCE, nd, p. 583.

7. Council on Environmental Quality. Environmental
    Quality--1979, U.S. Government Printing Office,
    Washington, D.C., 1980, p. 136.

8. Tourbier, Joachim. The Christina Basin: The Protection
    of Water Resources as a Basis for Planning in Developing
    Areas, Water Resources Center, University of Delaware,
    Newark, Del, 1973.

9. Tourbier, J. Toby and Richard Westmacott. Water
    Resources Protection Technology: A Handbook of Measures
    to Protect Water Resources in Land Development, ULI-the
    Urban Land Institute, Washington, D.C., 1981, p. 5.

10. Council on Environmental Quality. Environmental
    Quality -1978, U.S. Government Printing Office,
    Washington, D.C., 1979, p. 119.

11. Tourbier, J. Toby and Richard Westmacott. Water
    Resources Protection Technology: A Handbook of Measures
    to Protect Water Resources in Land Development, ULI-the
    Urban Land Institute, Washington, D.C., 1981, p. 5.

12. Wildrick, John T., Kurt Kuhn, and Waldon R. Kerns. Urban Water Runoff and Water Quality Control, Virginia Water Resources Center, Virginia Polytechnic Institute and State University, Blacksburg, Va., December 1976, p. 1.

13. U.S. Environmental Protection Agency. Water Quality Management Planning for Urban Runoff, U.S. Environmental Protection Agency, Washington, D.C., December 1974, p. 1.

14. Tourbier, J. Toby and Richard Westmacott. Water Resources Protection Technology: A Handbook of Measures to Protect Water Resources in Land Development, ULI-the Urban Land Institute, Washington, D.C., 1981, p. 10.

15. Castorina, Anthony B. "Surveillance and Monitoring Program for Connecticut Public Water Supply Watersheds," in Drinking Water Quality Enhancement Through Source Protection, Robert B. Pojasek, ed., Ann Arbor Science Publishers, Inc., Ann Arbor, Mich., 1977, p. 141.

16. Council on Environmental Quality. Environmental Quality--1980, U.S. Government Printing Office, Washington, D.C., 1981, p. 131.

17. A Manual for the Evaluation of a State Drinking Water Supply Program, U.S. Environmental Protection Agency, Water Supply Division, Washington, D.C., 1974.

18. Manual for Evaluating Public Water Supplies, U.S. Environmental Protection Agency, Office of Water Programs, PHS Publication 1820, U.S. Government Printing Office, Washington, D.C., Reprinted 1971.

19. Surveillance of Drinking Water Quality, WHO Monograph Series, Number 63, 1976.

20. Gerber, Robert G. "Land Use Controls in Watershed and Aquifer Recharge Areas," Journal of the Maine Water Utilities Association, Vol. 49, No. 6 (1973).

21. Van Nierop, Emannuel Theodorus. A Framework for the Multiple Use of Municipal Water Supply Areas, Cornell University Water Resources Center, Ithaca, N.Y., nd.

22. Hopkins, E.S. "Discussion Paper on D.S. Gilly, 'Management of Small Watersheds,'" Journal of the American Water Works Association, Vol. 45(1953), pp. 453-456.

23. Brown, C.B. "Erosion Control on Watershed Lands," Journal of the American Water Works Association, Vol. 38 (1946), pp. 1127-1137.

24. Gerber, Robert G. "Land Use Controls in Watershed and Aquifer Recharge Areas," Journal of the Maine Water Utilities Association, Vol. 49, No. 6 (1973).

25. Coughlin, Robert E. and Thomas R. Hammer. Stream Quality Preservation Through Planned Urban Development, U.S. Government Printing Office, Washington, D.C., 1973, p. 9.

26. Einsweiler, Robert C., Robert H. Freilich, Michael E. Gleeson, and Martin Leitner. The Design of State, Regional and Local Development Management Systems, Vol. I, Hubert H. Humphrey Institute of Public Affairs, University of Minnesota, Minneapolis, Minn., March 1978.

27. Hammer, Thomas R. (Betz Environmental Engineers, Inc.). Planning Methodologies for Analysis of Land Use/Water Quality Relationships, U.S. Environmental Protection Agency, Washington, D.C., October 1976.

28. Kusler, Jon. Regulating Sensitive Lands, Ballinger Publishing Company, Cambridge, Mass., 1980, p. 19.

29. Coughlin, Robert E. and Thomas R. Hammer. Stream Quality Preservation Through Planned Urban Development, U.S. Government Printing Office, Washington, D.C., 1973, p. 14.

30. U.S. Comptroller General. Water Quality Planning is Not Comprehensive and May Not Be Effective for Many Years, U.S. General Accounting Office, Washington, D.C., 1978.

31. Goldrosen, John. "The Role of Section 208 Planning in Protecting Drinking Water Sources," in Pojasek, 1977, pp. 42-43.

32. Kusler, Jon. Regulating Sensitive Lands, Ballinger Publishing Company, Cambridge, Mass., 1980, p. 19.

33. Aspen Systems Corporation. Compilation of Federal, State and Local Laws Controlling Nonpoint Pollutants: An Analysis of Laws Affecting Agriculture, Construction, Mining and Silviculture Activity, U.S. Government Printing Office, Washington, D.C., September 1975, pp. 71, 78.

34. Davey, William B. Conservation Districts and 208 Water
     Quality Management: Non-Point Source Identification and
     Assessment, Selection of Best Management Practices,
     Management Agencies, Regulatory Programs, Prepared for
     the U.S. Environmental Protection Agency by the National
     Association of Conservation Districts, National
     Technical Information Service, Springfield, Va., June
     1977.

CHAPTER III

THE PLANNING PROCESS AND PROBLEM ANALYSIS

Survey results reported in Chapter II indicate that in the United States over 85 percent of the watersheds feeding public water supplies are subject to some form of development. These results also indicate that development within those watersheds is followed by degradation in water quality from a variety of pollution sources.

Local governments have been operating a variety of measures that have mitigated to some extent the adverse effects of development, but few water system managers judged those measures to be very effective. They attributed the ineffectiveness of programs to the ad hoc manner in which management techniques are adopted and implemented. Only one in five local governments in the survey had adopted watershed protection measures as a part of a comprehensive program in which water quality management was a specific goal. However, respondents to the questionnaire found that significant improvements in water quality could be achieved with very effective watershed management programs.

In the remainder of the Guidebook we outline a process for developing a watershed management program to protect drinking water supplies. That process consists of five basic steps. They are:

1. Problem Analysis—The process of identifying existing and emerging water quality problems, estimating their potential magnitude, and evaluating their importance.

2. Direction Setting—The establishment of goals and their reduction to operational objectives.

3. Specification of Land Use and Physical Control Measures—The selection of cost-effective methods which when applied in the field will reduce pollution loads, modify pollutants, or regulate their discharge into water supplies.

81

4. Design and Implementation of Management
    Programs—The selection and implementation of
    incentives and organizational arrangements
    necessary to induce those who use the
    watersheds to adopt, operate, and maintain
    physical controls.

5. Monitoring and Evaluation—The process of
    auditing program effectiveness and identifying
    needs for adjustment.

In this chapter we examine the first of these steps,
problem analysis, in detail. Chapter IV addresses the
process of setting directions—reducing general goals to
operational objectives and principles, deciding on targets
for the management program, and formulating an appropriate
pattern of uses in the watershed. A wide variety of
intervention measures is discussed in Chapter V, and a number
of factors to be considered in formulating management
programs are discussed in Chapter VI. The last step, that of
monitoring and evaluation, is the subject of Chapter VII.

INTRODUCTION TO PROBLEM ANALYSIS

Problem analysis in watershed management may be viewed
as a special case of environmental risk assessment. Within
the past few years a large volume of literature pertinent to
that problem has emerged.[1] Its basic purposes are:

1. To identify qualitatively those pollutants and
    their sources which may, under existing
    policies and practices, cause significant
    adverse effects on public health, economic
    development, and environmental quality.

2. To estimate the likely magnitudes and
    frequencies of pollution and environmental
    degradation in as quantitative a manner as is
    necessary and possible.

3. To evaluate the relative importance of these
    effects as a basis for guiding the formulation
    of management strategies.

An inherent characteristic in the analysis is uncertainty.
There are uncertainties about events that initiate pollution
and environmental damage, uncertainties about the fate of
materials that are released to the environment, and

uncertainties about the effects that those materials will have on natural systems and upon humans. These uncertainties make problem analysis an imprecise art and complicate its conduct. Nonetheless, analysis is a necessary prerequisite to the design of management strategies, and uncertainty should be explicitly recognized in the process.

Figure I-1 (page 5) illustrates the elements that must be analyzed in evaluating surface water supplies. Most important among the elements usually included in the analysis are: (1) activity patterns on the watershed; (2) on-site pollution control measures; (3) natural systems through which pollutants pass before reaching major streams or reservoirs; (4) the aquatic environment through which pollutants are transported to water supply intakes; and (5) water treatment and distribution systems.

An analysis of the system to relate water quality management strategies to health effects in the exposed population requires, at least in concept, identification of relationships among the individual elements listed above. Those relationships, illustrated in Figure III-1, include:

1. Relationships between land use controls and activity patterns.

2. Relationships between on-site controls and pollutant loads released to the watershed.

3. Relationships between pollutants discharged and released in the watershed and pollutant loads delivered to receiving streams and reservoirs.

4. Relationships between pollutant loads entering receiving bodies and ambient concentrations of pollutants at water supply intakes.

5. Relationships between ambient concentrations of pollutants at the intake and the quality of drinking water.

6. Dose-response relationships relating quality of drinking water and the incidence of disease.

If these relationships are established, then a relatively simple model can be constructed to relate health effects to other elements in the system. One such model is given as follows:

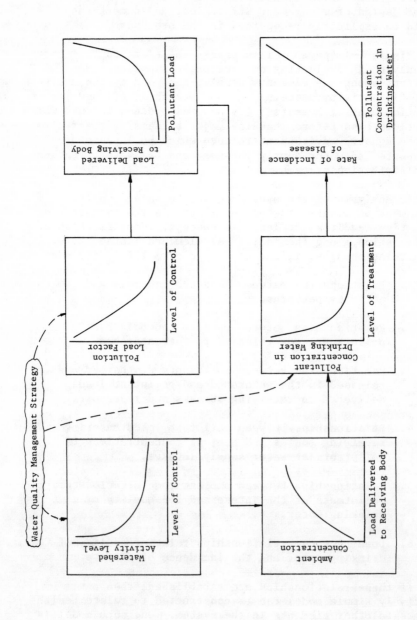

Figure III-1. Relationships Among Elements in Drinking Water Quality Management Systems

$$(1-e_p) \cdot S \cdot Q \cdot (1-e_w) \cdot R \cdot P$$

where A = Activity level in acres

L = Pollutant load in pounds per acre per year

e_p = Fraction of pollutants removed in on-site controls

S = Pollutant delivery ratio, the fraction of pollutants that pass on-site controls that are delivered to the receiving body

Q = Water quality transfer coefficient, i.e., the concentration of the pollutant (or its product) in the receiving water in mg/l per pound per year of pollutant added to that water

e_w = Fraction of pollutant removed in the water treatment process

R = Slope of the dose-response curve in number of cases of disease per mission persons exposed to the pollutant per mg/l of the pollutant in the finished water

P = Population in millions

D = Resulting health effect in cases of a disease

While this model is dimensionally correct and may be appropriate for some pollutants, it is highly simplified. It considers only a single activity; it assumes that the watershed transport system can be summarized by a simple constant, namely a pollutant delivery ratio (analogous to the sediment delivery ratio for soil transport); it implies a simple linear transformation of loads into ambient water quality; it assumes a simple proportional mechanism for the effect of water treatment; and it assumes a linear no-threshold dose-response relationship for health effects. No interactions with other substances with which the pollutant is in contact are assumed, nor is any synergistic or inhibitor behavior assumed in the dose-response relationship.

In reality relationships among elements in the system are much more complex than indicated by the simple two-dimensional curves shown in Figure III-1. The sets of factors are not independent of one another. Activity patterns on the watershed will depend in part upon water

quality and land use management strategies; natural processes
in the watershed will depend upon the nature of activity
patterns, and so on. Another source of complexity is the
dynamic behavior of watersheds requiring the introduction of
the temporal characteristics of discharges and storm events
to fully describe the system. Yet another source of complexity is the apparent random behavior of some variables,
most notably rainfall, that require probabilistic descriptions of certain variables and their relationships.

ELEMENTS OF ANALYSIS

Because of the complexities and uncertainties that one
is likely to encounter in any specific watershed, we refrain
here from advocating any particular quantitative model for
analysis. Rather we focus on those basic elements that
should be considered in the analysis. Later we will discuss
a sequential decision process for executing those basic
steps, and we will cite some analytical aids for carrying out
critical steps in the analysis at appropriate levels of
detail.

Figure III-2 is a schematic of the basic steps in
problem analysis. It is initiated with a scoping process in
which at least tentative bounds on the analysis are established. That step is followed by a descriptive analysis of
the natural features and processes of the watershed and of
the existing and probable future activity patterns that may
affect water quality. Given the complexity and uncertainty
associated with the analysis, the most important information
may be generated in the activity analysis, and in succeeding
steps where pollutants are identified and quantified and
where the effectiveness of post-generation pollution control
measures is assessed. The steps of tracing the fates of
pollutants through the watershed, estimating their impacts on
receiving streams, tracing their fate through water treatment
processes, and finally assessing their effects on the consuming population are more difficult and uncertain, but those
steps can hardly be ignored.

1. Scoping

Scoping is a process in which fundamental decisions are
made as to the bounds of analysis. At a minimum it must
specify: (1) the geographical coverage and period of time
relevant to the analysis; (2) guides as to what types of
pollution are relevant, including at least a tentative
inventory of pollutants and the activities that generate

86

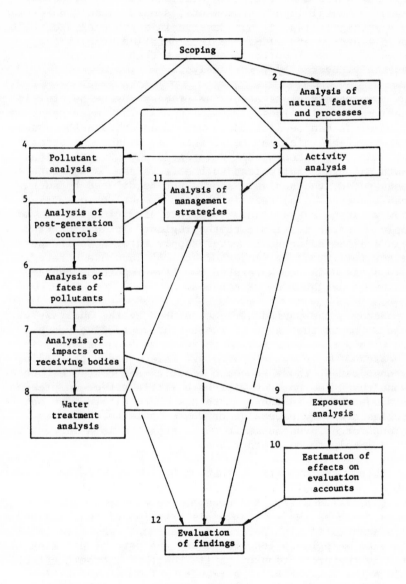

Figure III-2. Steps in Watershed Management
Problem Analysis

them; and (3) guides for evaluation, including the specification of social, economic, and environmental factors to be
considered in determining the relative importance of threats
to water supplies and the standards by which those threats
are to be judged. As the Council on Environmental Quality
outlined in its most recent regulations for the preparation
of environmental impact statements, scoping should determine
the significant issues that deserve in-depth analysis.[2] It
should identify and eliminate from detailed study those
issues of lesser importance and those that have been
adequately covered by prior analysis.

Some observations about the process may be helpful in
establishing guidelines. First, in most instances geographical coverage can be limited to water supply watersheds and
adjacent areas which may be affected by management strategies. Actions taken within watersheds may affect regional
growth patterns, but unless such effects can be adequately
demonstrated, the analysis should not be diluted by such
broad concerns. The test is whether or not actions taken in
a few watersheds are likely to affect significantly the total
supply of water available to the region. It is important to
include areas adjacent to water supply watersheds if there is
the possibility of spillover effects of land use strategies.
Actions taken in one watershed may have important consequences on neighboring watersheds.

Second, the period of time covered by the analysis will
probably be in the range of twenty to twenty-five years.
Even within that time period there is much uncertainty in
forecasts of development patterns. Also, under prevailing
discount rates, those economic benefits and costs that occur
beyond that time horizon will have little effect on present
decisions. Only in those instances where alternative water
supplies are very expensive or where it is necessary to
protect large investments in water supplies is a longer time
period likely to be appropriate.

Third, the objectives that should be considered in the
scoping process will usually include public health,
environmental quality, and economic development. Other
values, such as the equitable distribution of benefits and
costs associated with watershed management, may also be
relevant. A primary concern in any analysis of drinking
water is the protection of public health. Economic effects
include the loss of storage resulting from sedimentation in
reservoirs, loss of water supplies that are degraded in
quality below acceptable levels, losses in agricultural
production resulting from controls on land use and production

practices, and diseconomies that may arise in locational decisions for residential, commercial, and industrial activities. Environmental quality values are reflected in part by any reductions in aesthetic values and the quality of recreational experiences.

2. Analysis of Natural Features and Processes

If the information has not been compiled previously, it will be necessary to describe the natural physical features of the watershed and to identify significant natural systems. This analysis will consist largely of a compilation of descriptive information (tables, graphs, and maps) about the area, including:

1. Topography
2. Soils and geology
3. Vegetative covers
4. Precipitation and streamflow patterns
5. Wetlands and floodplains
6. Other relevant features and processes

3. Activity Analysis

Information about natural attributes of the watershed must be accompanied by an analysis of human activity patterns as they exist at present and as they may reasonably be expected to exist over the specified time horizon. This analysis should include, as appropriate, the following categories of activities:

1. Transportation (highways, railroads, pipelines, etc.)
2. Residential
3. Commercial
4. Industrial
5. Agricultural
6. Silvicultural
7. Other major activities

Relevant descriptions may include: (1) the location of facilities; (2) locations and areal extent of activities; (3) levels of activities as measured by indicators such as traffic flows, rates of production, number of employees, population, and numbers of housing units; and (4) technologies for producing and consuming goods. Levels of aggregation for these descriptions will depend on site-specific information. In some cases aggregate indicators may suffice; in others where individual activities or actors may dominate or constitute a significant threat, analysis of individual firms, parcels of land, segments of transportation arteries, and development units may be necessary.

Analysis of future activity patterns will require the formulation of several scenarios covering a range of possible conditions. Among the scenarios to be constructed may be: (1) the worst case conditions; (2) a trend-based condition; and (3) a condition that would reflect the imposition of strict water-quality related development controls.

4. Pollutant Analysis

Given the analysis of natural characteristics and the nature of activity patterns within a watershed, water quality specialists should be able to <u>identify</u> a probable inventory of pollutants being generated by the activities. By consulting the appropriate literature, secondary <u>estimates of probable rate</u>s at which those pollutants are being generated may also be constructed through comparative analyses of other watersheds on which primary field data have been collected previously.

Inventories constructed using secondary data sources will contain uncertainties about both the <u>existence</u> of selected pollutants and <u>rates</u> at which those materials are or will be generated within a watershed. These uncertainties can be reduced by surveys involving the collection and analysis of field samples from pollutant generating activities, including: (1) those that generate pollutants in a more or less continuous pattern; (2) those that generate pollutants only during storm events; and (3) those from which pollutants would be generated only in case of an accident or other unforeseen event. The decision to purchase primary field information should be made after weighing the probable consequences of the several pollutants, levels of uncertainty about their existence and rates, and extent to which that uncertainty could be reduced against the cost of obtaining the information. Of course, primary information about future activities can be obtained only after those activities commence.

No attempt is made here to construct exhaustive lists of potential pollutants and their sources. Some of the most commonly encountered contaminants, their effects, and their sources are discussed in Chapter I. Figure III-3 is a pollutant-by-source matrix that may be useful as a checklist

TYPE OF PROBLEM

SOURCE	Biological	Sedimentation	Oils/Grease	Metals	Pesticides	Other Hazardous Chemicals	Nutrients	Acids/Gases	Radiation
Residential									
Commercial									
Industrial									
Transportation (highway, rail, pipelines)									
Agricultural									
Silvicultural									
Mining									
Surface runoff									
Groundwater									
Atmosphere									

Figure III-3. Pollutant-by-Source Matrix

to identify important cause—effect linkages within a given
watershed.

5. Analysis of Post—Generation Controls

Some of the pollutants that are included in the inventory will have been identified in previous studies and
post—generation control systems will be in place. Those
systems must be identified and an analysis of their performance must be made to determine rates and temporal patterns
at which pollutants are released to the environment. Special
attention should be given to the consequences that would
follow failures in these controls.

6. Analysis of Fates of Pollutants

The study of the process by which pollutants are
transported to receiving waters should be made. This
includes an analysis of the transport and chemical and
biochemical transformations that occur, and concludes with
estimates of temporal and spatial patterns over which
pollutants are delivered to receiving bodies of water.

7. Analysis of Impacts on Receiving Bodies

Predictions must be made of ambient concentrations, the
entrapment of materials in benthic deposits, and concentrations of chemical and biochemical products. In this step the
analyst is coming to basic conclusions about water quality
impacts of activities as mitigated by post—generation controls and natural processes.

8. Water Treatment Analysis

The next step in the process consists of analyzing the
effects of both existing, and to a lesser extent at this
stage, alternative drinking water treatment processes on
pollutants and their products and vice versa, i.e., the
effects of those materials in treatment processes.

In steps six through eight the choice between reliance
on secondary sources of information and the acquisition of
primary field data is similar to that for identifying and
estimating rates at which pollutants are being generated.
Again it is a decision about the allocation of planning
resources and should be based on some assessment of the

anticipated value of additional information versus its costs. Information generated in the preceding elements of the problem analysis should result in estimates of the ambient concentrations of pollutants and their products and the concentrations of pollutants in drinking water. Those estimates should be accompanied by some indicators of the likelihood with which those concentrations will occur, and where possible, estimates of the duration and frequency of occurrences. That information then provides a basis for the next two elements of the analysis.

9. Exposure Analysis

Estimation of the numbers of persons exposed to identified hazards and the conditions of their exposure is made to obtain a basis for estimating health effects.

10. Estimation of Effects on Evaluation Accounts

Findings from the previous steps must be assessed on the basis of objectives established in the scoping step. In particular, the findings must be translated in terms of the exposure analysis into public health effects, the economic consequences of loss of storage and reductions in water quality, and environmental quality effects on streams and impoundments.

11. Analysis of Management Strategies

Existing strategies for water and related land management should be examined as an element in problem analysis. That examination should address the management strategies on three points: first, by describing the institutional structure for management; second, by describing the set of action-forcing measures employed by the institutions; and third, by describing the responses of activities to those measures as indicated by land use changes (or lack of change), investments in water quality controls, operating procedures, and materials used in production, consumption, and maintenance processes.

12. Evaluation of Findings

The final step in problem analysis is one of evaluating findings from earlier steps in the process. The issues to be resolved are:

93

1. Which pollutants, if any, are the most important ones to be addressed in formulating new or modified management strategies?

2. Which activities are the most important sources of those pollutants or otherwise pose the most important threats to the water supply?

3. Which elements of existing management strategies are either inadequate or most likely to fail to provide adequate protection?

4. What are the most important sources of uncertainties? Can they be reduced through field surveys? And, does the anticipated value of information from additional surveys justify the cost of obtaining that information?

EXECUTION OF THE PROBLEM ANALYSIS

Execution of all the elements of problem analysis in a comprehensive and detailed manner could overtax the financial resources and expertise available to many communities for this purpose. Such an effort may not be necessary to support some of the basic decisions about watershed protection. For this reason the analysis should properly be carried out sequentially at two or more levels of detail. The first pass would be at a reconnaissance level where readily available information from earlier studies and secondary sources would be used. At that level the analysis will be heavily dependent on expert opinion: (1) to translate activity patterns into pollutants and pollutant loads; (2) to estimate impacts on ambient water quality; and (3) to identify potential hazards. Second and subsequent passes through the analysis would be in greater detail, focused more narrowly on those problems or potential problems which were identified as being important in the first pass. Decisions to acquire additional field data or to analyze existing data more thoroughly would be implemented in those rounds.

Figure III-4 shows the sequence of decisions that must be made in the process. Two questions emerge from the first pass. First, has enough information been generated to drop one or more pollutants from the inventory of possible problems? Such a conclusion could be drawn under any of the following conditions:

1. Levels of activity that would generate that pollutant are sufficiently small.

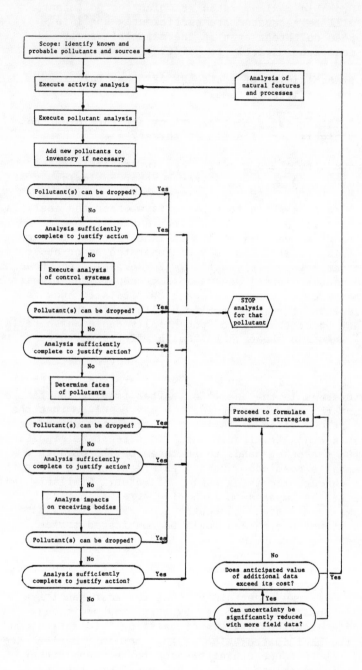

Figure III-4. Sequential Decision Process
for Execution of Problem Analysis

2. Probable maximum rates at which the pollutant
   will be generated are sufficiently small to
   pose no threat even if the material was
   delivered directly to points of human contact.

3. Probabilities of accidents or other failures
   are sufficiently small.

4. Existing management practices are sufficient to
   mitigate any significant threat.

If the answers to these question are in the affirmative, then
that pollutant can be dropped from further analysis.

The second question to explore is whether the analysis
is sufficient to justify action with respect to one or more
pollutants. For some pollutants it is sufficient to know
they are being generated or may be generated at or about
specified rates to justify action. For example, one does not
need to construct formal analyses of transport and transformation processes on some persistent chemicals to know that
they represent a threat to public health. For others
uncertainty about the fate of potentially dangerous materials
in a watershed and about their health effects may dictate the
adoption of controls as a precaution.

The same questions may arise after proceeding with
subsequent steps in the analysis. After tracing the fate of
a pollutant through a watershed, one may conclude that the
probable maximum rates of delivery to receiving streams or
impoundments are sufficiently small to present no significant
hazard. On the other hand, those rates may be of such a
magnitude as to conclude a hazard exists without tracing the
pollutant through receiving bodies of water. Similarly, with
estimates of the impacts on receiving streams in hand,
probable maximum ambient concentrations may be below some
threshold at which a threat would be considered to be
significant, or vice versa.

Finally, an analyst, knowing something of the uncertainties that are present in the estimates, must consider the
questions: (1) can uncertainty about the most sensitive
items of information be reduced by obtaining additional field
data or extending the analysis of existing information, and
(2) does the anticipated value of that new information exceed
its cost? If that uncertainty cannot be reduced, either
because it is associated with future events or because it is
a long-term research issue, then management strategies must
be based on available information with due regard for the
uncertainty. A similar course would be followed if the cost

of obtaining information outweighs its anticipated value. If
the uncertainty can be reduced in a cost-effective manner,
then it is prudent to obtain that information on subsequent
passes through the analysis.

MODELS TO AID IN PROBLEM ANALYSIS

A variety of analytical methods may be used to execute
the basic steps of analysis depicted in Figure III-2. Three
areas are particularly well suited to more formal modeling
methods, namely: (1) the construction of scenarios of
activity patterns that are likely to occur on the watersheds;
(2) the transport of materials across land surfaces from
their points of release to their points of entry into
receiving streams; and (3) their effects upon receiving
streams, particularly upon impoundments from which drinking
water supplies are taken.

Activity Models

Activities must be described as they exist at present
and as they may reasonably be anticipated to develop over the
time horizon. Analysis of present activities will consist of
descriptive statistics, maps, and other forms of representation. Anticipated future conditions will consist of some
form of projections or scenarios of activity patterns. To
construct those scenarios, one of several types of models
will probably be required, models which range in complexity
from simple trend projections to those which attempt to
simulate future market conditions.

Components of activity patterns that would normally be
included in the analysis are:

1. residential
2. commercial
3. industrial
4. agricultural
5. silvicultural
6. transportation

Others, such as mining and recreation, may be relevant to
particular watersheds.

Most likely the activities will be analyzed at two
levels of detail and areal coverage. At the more general
level activities will be described by areawide land use and
transportation analyses, showing spatial patterns and

97

supporting indicators of levels of activity within each
subarea for each category of activity. Where appropriate,
daily or seasonal variations in the levels of these
activities may be included. At the second level of detail
analysis of <u>individual units</u> of activity may be necessary,
where those units may consist of either: (1) individual
industrial firms, agricultural enterprises, mining
operations, or other unit which by itself may pose a threat
to the water supply; or (2) a representative sample of
activity units, such as households or small to moderate size
farms, which taken collectively may pose a threat.

     Scenarios of future activity patterns within water
supply watersheds and adjacent areas can be constructed using
either simple trend extrapolation models or more formal
models that simulate land development processes. The most
appropriate type of analysis is likely to be one that
extrapolates existing trends within subareas defined by
census tracts and enumeration districts. Careful analysis of
recent trends using spatial data from the census and
supporting data from tax maps and building permits is likely
to yield the most reliable insights into development patterns
that are occurring in or near the watershed. That
information, when combined with planned investments in
transportation facilities and other public and private
facilities is probably sufficient to forecast activity
patterns as far into the future as it is reasonable to
explore. However, simple trend extrapolations are probably
not sufficient. Those trends must be adjusted to reflect
constraints imposed by limited supplies of land, the
suitability of land resources to support various uses, and
public and private policies that encourage or restrict land
uses and practices. In constructing scenarios of development
patterns it is good practice to make those scenarios
consistent with projections of activities for the larger
geographical, political, or economic regions within which the
watershed is located. That larger region may be a county,
Standard Metropolitan Statistical Area, or some other unit
for which information about activity levels is routinely
collected and for which projections may be available.

     In addition to less formal methods that rely heavily on
trend extrapolation and ad hoc processing of information
about local land uses, there are more formal computer
simulation models for allocating activities and related land
uses to a set of geographical subareas. These include: (1)
models that are based on social physics concepts analogous to
gravity and potential models of natural science; (2)
econometric models calibrated to existing activity patterns;
and (3) models that attempt to simulate housing markets.

Details of these models may be found in a number of standard textbooks* as well as the primary references to which those books refer. All of these models require the use of digital computers, they require the preparation of extensive data bases, and they require a relatively high level of technical expertise in their construction and the interpretation of results. Unless such models have been previously developed by regional planning agencies it is unlikely that they should be used in watershed problem analysis.

Land Surface Models and Receiving Water Models for
    Pollutant Analysis

Once the patterns at which pollutants are released to the environment are known, the analytical problem is one of tracing their fates from those points in time and space where they enter the environment to those points of use at which measures of water quality are necessary. In reality that movement may be in a continuous flow, but for ease of analysis it is most often divided into two basic components, one tracing the flow over land surfaces and the second tracing flows in receiving bodies. In both components the pollutants are dispersed into the environment through a complex set of natural processes including those that transport materials and those that transform the states of matter and energy of the pollutants and their products.

A variety of models for exploring and explaining the flows of materials in each of those components have been developed. Those models generally fall into one of two categories within which there are many variations. The first category is referred to as statistical; it includes a range of models from simple estimates of input-output coefficients to multivariate models with time lags. The second category is analytical based on applications of the principles of conservation of mass and energy.

Statistical models encompass a wide variety of empirical methods for which few generalizations are possible. They do not explicitly represent the various transport and transportation processes at work in nature. Rather they tend to treat those processes as "black boxes" in which it is either impossible or unnecessary to know causal mechanisms that relate "input" or "independent" variables to "output" or "dependent" variables. Those variables are related empirically solely on the basis of field measurements of

---

*For example, see references 2 and 3.

inputs or outputs. In the simplest case a model may consist of a set of coefficients by which inputs are multiplied to estimate outputs. In more complex statistical models one or more output variables may be related to multiple input variables through linear or nonlinear regression equations with or without time lags among the several variables.

Analytical models are applications of the fundamental laws of conservation of mass or energy. Those laws provide basic accounting systems in which time averaged rates of the accumulation of mass or energy within a volumetric unit must be balanced by rates of inflow, outflow, and physical, chemical, and biological rates of transformation.

Temporal and spatial characteristics of either class of models must be specified early in the process. At least two attributes of the analysis must be made specific. First, an averaging time must be established. The averaging time is the basic unit of time over which variables are to be esti- mated. For example, phosphorous concentrations in a reser- voir may be stated as an annual average, as a seasonal or daily average, or as instantaneous values. The choice of an appropriate time is highly dependent on the nature of the problem. If the problem is one of tracing the flow of nutrients in a reservoir, monthly or seasonal units may be appropriate. If, on the other hand, the problem is one of analyzing impacts from an accidental spill, much shorter time intervals are appropriate, i.e., hours or days. The second and somewhat related choice is between event oriented and continuous models. Event oriented models are those that trace the effect of a single event, such as a single storm or an accidental spill under a given set of environmental con- ditions. Continuous models are those that account for the flows of materials over many successive time intervals span- ning a long time horizon of several seasons or even years.

Land Surface Models

While mass and energy balance models for analyzing the flow of stormwater runoff over land surfaces have been developed extensively in the literature, few have attempted to incorporate the flow of pollutants in those models. More complex models, such as EPA's Storm Water Management Model, combine models of overland flow with those for flow in channels (street gutters and storm sewers) to represent complex patterns of flow in urban areas. Some of those do include flows of pollutants, but even in those models the analysis is quite limited. Pollutants are traced only after they reach a channel; they are assumed to be transported by

advective processes; and no chemical or biochemical transformations are considered. These are known to be overly
simplistic; the models are not yet capable of handling the
complex and only partially understood bed-load and other
transport phenomena that are known to occur. Huber and
Heaney[4] have recently reviewed the population of mass balance
models for land surface runoff. A selected list of those
models is shown in Table III-1. Models that appear in the
list were chosen because: (1) they include water quality in
the analysis; (2) they are applicable to nonurban as well as
urbanizing watersheds; and (3) there exists some experience
with their application, either extensive experience reported
by the authors or experience reported by others. The first
two of those models are relatively simple to use but must be
viewed as first-cut analyses. Others in the list are more
sophisticated, requiring the use of digital computers and a
higher level of expertise in the selection of input data,
programming, and interpretation of outputs. The Hydrocomp
Simulation Program, Agricultural Runoff Management Model,
Agricultural Runoff Model, and Storm Water Management Model
are capable of representing single storm events; the others
are continuous simulation models representing many storm
events and intervening dry periods.

Given the relative newness of these models, the levels
of expertise required to calibrate them and interpret results
from them, and the questionable quality of those outputs, it
is not surprising that most estimates of the flows of
pollutants from land surfaces are made using relatively
simple statistical models. Probably the most widely applied
of these are based on an extension of methods for the
analysis of soil erosion and sedimentation. In particular
they are based on an extension of the Universal Soil Loss
equation. That equation is used to estimate "gross" erosion
rates of sheet erosion; i.e., the amount of soil lost from a
given type of land surface over a given period of time. It
does not estimate the proportion of that load that enters a
receiving stream. That estimate is obtained by multiplying
gross erosion rates by the sediment delivery ratio, another
statistically derived quantity which can be adjusted to
reflect the influences of:[5]

1. Sizes of drainage areas

2. Lengths, slopes and the density of well-defined
   (incised) channels

3. Lengths and slopes of the watershed

4. Precipitation and runoff characteristics

TABLE III-1

MASS BALANCE MODELS FOR LAND SURFACE RUNOFF

Name	Reference
Midwest Research Loading Functions	McElroy et al., 1976
Environmental Pollution Assessment: Erosion, Sedimentation and Rural Runoff Model	True, 1976
Hydrocomp Simulation Program	Hydrocomp, Inc., 1976
Pesticide Transport and Runoff Model	Crawford and Donigian, 1973
Agricultural Runoff Management Model	Donigian et al., 1977
Nonpoint Source Model	Donigian and Crawford, 1976
Terrestrial Ecosystem and Hydrology Model	Patterson et al., 1974
Simulation of Contaminant Reactions and Movement	Adams and Kurisu, 1976
Agricultural Chemical Transport Model	Frere et al., 1975
Storage, Treatment, Overflow and Runoff Model	Hydrological Engineering Center, 1977
Agricultural Runoff Model	Roesner et al., 1975
Storm Water Management Model	Metcalf and Eddy, Inc., et al., 1971

Source: W.C. Huber and J.P. Heaney, "Analyzing Residuals Generation and Discharge from the Land Surface," in Analyzing Natural Systems, D.J. Basta and B.T. Bower, eds., Research Paper, Resources for the Future, Inc., Washington, D.C., 1982, Chapter 3.

Table III-1 (continued)
References for Table III-1

McElroy, A.D., S.Y. Chiu, J.W. Nebgen, A. Aleti, and F.W. Bennett. 1976. Loading Functions for Assessment of Water Pollution from Nonpoint Sources, Report No. EPA-600/76-151, prepared for Environmental Protection Agency, Washington, D.C., May, also available from NTIS PB-253 325.

True, H.A. 1976. "Planning Models for Non-Point Runoff Assessment," in W.R. Ott, ed., Proceedings of the Conference on Environmental Modeling and Simulation, Report no. EPA-600/9-76-016 (Cincinnati, Ohio, Environmental Protection Agency, July), pp. 74-76, also available from NTIS PB-257 142.

Hydrocomp, Inc. 1976. Hydrocomp Simulation Programming and Operations Manual (4th ed., Palo Alto, Calif.).

Crawford, N.H. and A.S. Donigian, Jr. 1973. Pesticide Transport and Runoff Model for Agricultural Lands, Report No. EPA-660/2-74-013, prepared for Environmental Protection Agency, Washington, D.C., December, also available from NTIS PB-235 723.

Donigian, A.D., Jr., D.C. Beyerlein, H.H. Davis, and N.H. Crawford. 1977. Agricultural Runoff Management (ARM) Model--Version II: Refinement and Testing, Report No. EPA-600/3-77-098, prepared for Environmental Protection Agency, Athens, Ga., August.

Donigian, A.S., Jr., and N.H. Crawford. 1976. Modeling Nonpoint Pollution from Land Surface, Report No. EPA-600/3-76-983, prepared for Environmental Protection Agency, Athens, Ga., July, also available from NTIS PB-250 566.
Patterson, M.R. et al. 1974. A User's Manual for the Fortran IV Version of the Wisconsin Hydrologic Transport Model (Oak Ridge, Tenn., Oak Ridge National Laboratory).

Adams, R.T. and F.M. Kurisu. 1976. Simulation of Pesticides Movement on Small Agricultural Watershed, Report No. EPA-600/3-76-66 (Athens, Ga., Environmental Protection Agency).

Frere, M.H., C.A. Onstad, and H.N. Holtan. 1975. ACTMO--An Agricultural Chemical Transport Model, Publication No. ARS-H-3 (Hyattsville, Md., Agricultural Research Service, U.S. Department of Agriculture).

Table III-1 (continued)

Hydrologic Engineering Center (U.S. Army Corps of Engineers).
1977. Storage, Treatment, Overflow, Runoff Model, STORM
(Davis, California).

Roesner, L.A. et al. 1975. Agricultural Watershed Runoff
Model for the Iowa-Cedar River Basins, prepared for the
U.S. Environmental Protection Agency by Water Resources
Engineering Inc., Walnut Creek, Calif.

Metcalf and Eddy, Inc. et al. 1971. Storm Water Management
Model, Vol. I--Final Report, Report, No. 11024 DOC 07/71,
U.S. Environmental Protection Agency, Washington, D.C.,
also available from NTIS PB-203 289.

That method has been extended to other pollutants by assuming
that the chemical composition of sediment particles remains
constant from the points of erosion to the points where they
enter receiving streams. Thus, delivery rates of nutrients,
heavy metals, and other constituents can be obtained by
multiplying the rate of sediment delivery to the stream by
the pollutant-to-sediment ratio for eroded soils. That is
the approach taken in the Midwest Research Loading Functions,
the first model listed in Table III-1.

Receiving Water Models

Once the flows of materials have been traced from their
points of origin within a watershed to their entry into
receiving streams, the task becomes one of tracing their
flows and effects on streams and impoundments fed by those
streams.

Building upon early work by Streeter and Phelps[6] and
later developments by Thomas[7], O'Conner,[8] and others, a large
array of water quality models for streams, lakes, estuaries,
and ocean outfalls have been formulated, programmed, and
applied to various problem settings over the past decade.
Hinson and Basta[9] have recenctly reviewed a large number of
such models. Of particular interest here are those for
estimating effects on reservoirs, some of which are listed in
Table III-2. Among various elements that set of models
accounts for are: (1) transport processes, principally
advocation, diffusion, and sedimentation; (2) gas transfer,
especially air-water interchanges of oxygen; (3) biochemical
oxidation processes; and (4) ecological processes, including

TABLE III-2

SELECTED MODELS FOR ESTIMATING EFFECTS ON RESERVOIRS

Name	Reference
Simplified Stream Model	Hydroscience,Inc., 1971
Deep Reservoir Model	Water Resources Engineers, Inc., 1967
Lake Ecological Model	Finnemore and Shepherd, 1974
Reservoir Water Quality Model	Duke, 1974
Hydrocomp Simulation Program	Hydrocomp, Inc., 1972
Water Quality in River-Reservoir System	Hydrologic Engineering Center, 1977

Source: M.O. Hinson and D.J. Basta, "Analyzing Surface Water
Receiving Waters," in Analyzing Natural Systems,
D.J. Basta and B.T. Bower, eds., Research Paper,
Resources for the Future, Inc., Washington, D.C.,
1982, Chapter 4.

References for Table III-2

Hydroscience, Inc. 1971. Simplified Mathematical Modeling of
Water Quality, GPO 1971-444-367/392 (Washington, D.C., U.S.
Environmental Protection Agency, Office of Water Programs).

Water Resources Engineers, Inc. 1967. Prediction of Thermal
Energy in Streams and Reservoirs (Sacramento, Calif.,
California Department of Game and Fish).

Finnemore, E.J. and J.L. Shepherd. 1974. Spokane River
Basin Model Project: Volume I -Final Report (Seattle,
Wash., Environmental Protection Agency, Region X, Air and
Water Division).

Table III-1 (continued)

Duke, J.H., Jr. 1974. Computer Program Documentation for the Reservoir Water Quality Model EPARES (Washington, D.C., Environmental Protection Agency, Systems Development Branch, January).

Hydrocomp, Inc. 1972. Hydrocomp Simulation Programming: Mathematical Model of Water Quality Indices in Rivers and Impoundments (Palo Alto, Calif.).

Hydrologic Engineering Center (U.S. Army Corps of Engineers). 1977. Storage, Treatment, Overflow, Runoff Model, STORM (Davis, California).

photosynthetic production, growth and respiration of zooplankton, and processes involving organisms in higher trophic levels. As a set the models are applicable to a broad range of pollutants, including: (1) bacterial populations; (2) dissolved oxygen demanding materials; (3) nutrients and algal growth; (4) sediment; and (5) toxic substances. Most are rather sophisticated models requiring a substantial amount of programming capability and experience in applying and interpreting the results of computer simulation models for water quality.

Another class of models that have been widely applied to the analysis of reservoirs threatened by eutrophication are those developed by Vollenweider and Dillon,[10] and Larsen and Mercier.[11] They were developed especially to classify and predict trophic states of lakes and reservoirs as a function of phosphorous loads, detention times, and phosphorous retention coefficients. They are based on simple steady-state, one compartment representations of lakes and reservoirs. As such they are best used as coarse indicators of probable trophic states of those water bodies. A related analysis is the work by Shannon and Brezonik[12] which relates, through regression analysis, a trophic state indicator to loads of nitrogen and phosphorous on lakes and impoundments.

While there has been a large volume of work in developing and calibrating a variety of models for estimating the runoff of pollutants from watersheds and the effects of those pollutants on receiving bodies, their predictive powers in the absence of collaborating field data remain limited. They are poor substitutes for direct field measurements of materials as they are released into the environment, as they enter receiving streams, and as their effects are measured at drinking water intakes. A real limitation in their application is the proper representation of the combination of causal events that release materials to the environment, of pathways between points of release and transport into drinking water supplies, and the accumulation of materials in drinking water impoundments. Their results may provide useful insights into the formulation of management strategies, but they cannot be used as reliable predictors in the absence of primary field data.

SUMMARY

A process for developing a watershed management program was outlined at the beginning of this chapter:

1. Problem analysis

2. Formulating basic goals, principles, and use patterns for the watershed

3. Selecting physical control strategies

4. Designing and implementing a program of intervention measures and organizational arrangements

5. Monitoring and evaluation

The remainder of the chapter discussed the problem analysis step in the process. After describing watershed problem analysis as a special case of environmental risk assessment, the major elements of the analysis were outlined and a multiple-pass strategy for circulating through the elements was recommended. Modeling methods for several of the elements—activity analysis and analysis of pollutant transport and transformation over land and in water bodies—were described and assessed.

In the following chapter we focus on formulating the basic dimensions of a watershed management program. They include: (1) obtaining consensus on goals and management principles; (2) deciding on the target pollutant, sources, and hydrologic process to be influenced; and (3) specifying appropriate land and water uses in the watershed and the physical controls necessary to ensure an adequate and safe yield of raw water.

REFERENCES

1. Moreau, David H. Quantitative Assessments of Health Risks by Selected Federal Regulatory Agencies, prepared for the Office of Air Quality Planning and Standards, U.S. Environmental Protection Agency, Research Triangle Park, N.C., October 1980.

2. Reif, B. Models in Urban and Regional Planning, Intertext Educational Publishers, New York, N.Y., 1973.

3. Batty, M. Urban Modelling, Cambridge University Press, Cambridge, Mass., 1976.

4. Huber, W.C. and J.P. Heaney. "Analyzing Residuals Generation and Discharge from the Land Surface," in Analyzing Natural Systems, D.J. Basta and B.T. Bower, eds., Research Paper, Resources for the Future, Inc., Washington, D.C., 1982, Chapter 3.

5. Gottschalk, C.C. "Sedimentation," in Handbook of Applied Hydrology, V.T. Chow, ed., McGraw-Hill, New York, N.Y., 1964, Chapter 17.

6. Streeter, H.W. and E.B. Phelps. A Study of the Pollution and Natural Purification of the Ohio River, Public Health Bulletin 146, U.S. Public Health Service, Washington, D.C., 1925.

7. Thomas, H.A. "Pollution Load Capacity of Streams," Water and Sewage Works, Vol. 95 (1948), p. 407.

8. O'Connor, D.J. "Oxygen Balance of an Estuary," Journal of Sanitary Engineering Division, Proceedings of the American Society of Civil Engineers, Vol. 86, No. SA3 (May 1960), pp. 35-55.

9. Hinson, M.O. and D.J. Basta. "Analyzing Surface Water Receiving Bodies," in Analyzing Natural Systems, D.J. Basta and B.T. Bower, eds., Research Paper, Resources for the Future, Inc., Washington, D.C., 1982, Chapter 4.

10. Vollenweider, R.A. and D.J. Dillon. The Applications of the Phosphorous Load Concept to Eutrophication Research, prepared for the Associate Committee on Scientific Criteria for Environmental Quality, Burlington, Ontario, June 1974.

11. Larson, D.P. and H.T. Mercier. "Lake Phosphorous Loading Graphs: An Alternative," U.S. Environmental Protection Agency National Eutrophication Survey Working Paper No. 174, July 1975.

12. Shannon, E. E. and P.L. Brezonik. "Relationships Between Trophic State and Nitrogen and Phosphorous Loading Rates," Environmental Science and Technology, Vol. 6, No. 8 (1972), pp. 719-725.

CHAPTER IV

FORMULATING BASIC DIRECTIONS AND DIMENSIONS
OF THE WATERSHED MANAGEMENT PROGRAM

"A problem well stated is a problem half solved." But,
we might add, only half solved. Without detracting from the
value of a sound initial analysis, the task remains to formulate technically sound solutions, choose from among them, and
adapt and implement a program of action in the field.

The next three chapters address various aspects of the
formulation-and-implementation problem. This chapter focuses
on formulating the basic dimensions of a watershed management
program. These include: (1) specification of the primary
goals as well as an accompanying set of supplementary goals,
constraints, and solution criteria; (2) selection of target
pollutants, target sources, target locations, and target
hydrologic processes based on the earlier problem analysis
described in Chapter III and on the goal-constraint structure; and (3) specification of an appropriate combination of
land uses and practices in the watershed to ensure adequate
water quality and yield.

Chapter V, the following chapter, then outlines the
range of intervention measures that might be applied to
attain the basic solution framework outlined here. The
measures constitute potential action-forcing tools and
techniques for a specific management program. The third of
these chapters on program design and implementation, Chapter
VI, then addresses the heart of the matter—the formulation
of the actual watershed management program—that combination
of actions to be undertaken by the water system and
cooperating governments to protect raw water yield and
quality. That chapter describes the major issues involved
and the major choices to be made and discusses a process for
designing and implementing a program of specific actions.

DIRECTION-SETTING: GOALS, OBJECTIVES, CONSTRAINTS, SOLUTION
    CRITERIA, PRINCIPLES AND STANDARDS

Problem analysis, as discussed in Chapter III, constitutes a preliminary step in direction-setting. The diagnosis

111

of current and projected problem conditions and the prognosis of water supply needs are essential first steps in narrowing the search and deciding what to do. In addition to establishing existing and projected water quality and yield conditions, the problem analysis also suggests, as indicated in Chapter III:

1. Current and potential pollutants that are most affecting raw water quality and yield or are projected to do so.

2. Activities and locations that are the most important sources of those pollutants or that otherwise pose the most significant threats to the water supply.

3. Elements of existing management strategies that are inadequate, and why.

4. Transport and transformation processes that are most significant in determining the amounts and types of sediments and pollutants reaching the impoundment.

5. The institutional framework of factors that affect inclination and capability to implement a watershed management program (see Chapter VI).

All of these outputs help establish guidelines for solution searching.

Once these problems and needs have been systematically investigated in the problem analysis stage, however, it is important to go several steps further. First, the direction-setting process should achieve sufficient clarity and commitment about the overriding goal of water supply quality and yield and any other public interests that are to be sought in dealing with the water supply problem. This commitment to and clarity in basic purposes of a watershed management program is required of both the technical personnel working on the program and the appointed and elected officials who will be adopting any future management measures. Second, the direction-setting process should also include determination of and commitment to qualities of a management program that make the management program efficient, equitable, and feasible. Third, direction-setting involves the reduction of general goals and constraints into more explicit and useable guidelines. These guidelines take the form of (1) solution search principles and standards and (2) choice principles

112

and standards.

Determination of Goals

An early step in devising watershed management programs
should be to determine the goals which the program is to
achieve and to obtain a commitment to them among major
participants in the management process. As a recent study on
community water resources planning concludes,[1]

> A public consensus or, at a minimum, a clear
> statement of views, on the nature and magnitude of
> local resources problems and goals forms a basis
> for sound management and planning. In Stoughton,
> as in many other communities, there was no clear
> statement of problems or needs of the town to help
> guide the action of the Selectmen, the Town
> Engineer, or other Town Agencies, until the
> formulation of the Water Advisory Committee... A
> publicly accepted list of priorities would have
> helped the town to avoid future crises.

The primary goal on which commitment must be reached is
the assurance of a high quality water supply of adequate
yield. By water quality we generally mean how safe it is to
drink--the most important attribute of quality--as well as
how it smells, how it looks, and how it tastes. This general
goal of water quality and yield may be expressed in a variety
of ways, for example:

. ensuring an adequate supply of raw water...and keeping
  this supply pure (in the face of) unsewered development
  that is beginning to spread across reservoir drainage
  basins.[2]

. ...maintaining high water quality in the water
  supply reservoirs to ensure a continued supply of
  high quality water at a reasonable rate...(and)
  that benefits of and responsibilities for necessary
  actions be equitably shared by all parties.[3]

. meeting municipal needs for adequate supplies of
  fresh water in the most economically feasible and
  environmentally sound manner.[*]

---

*Adapted from reference 4.

. protection of...water (supply) resources...while accommodating economic activities...[4]

If not already made earlier in the planning process, the decision should be made at this point whether the goals of the watershed management program will focus on water supply yield and quality or be broadened to include other water resource and environmental quality interests, and even to comprehensive community interests. On the one hand, a focus on water supply yield and quality avoids at least temporarily the problem of trading off water supply goals against competing community goals and simplifies the planning problem. On the other hand, the more benefits promised by a proposed watershed management program, the more likely it is to be justified later in the evaluation and implementation stages and the more support it might generate from various interest groups and governments who must support the program or even be responsible for implementing portions of it. This is the nature of watershed management, which usually (1) involves governments whose electorates are not using the potable water but which have geographical jurisdiction over substantial portions of the watershed, and (2) involves restricting property rights of watershed landowners. Thus a wider range of goals is often considered in watershed management. In addition to water supply yield and quality goals, they include the following:

1. Open space

2. Protection of prime agricultural lands

3. Increased recreational opportunity

4. Enhancement of fish and wildlife resources

5. Preservation of significant water related ecological, geological, and cultural resources

6. Reduction of flood damage

7. Efficient use of public services

8. A particular urban development pattern, such as compact development

9. Economic development

This Guidebook advises expanding from the water supply protection goal only so far as is necessary to gain advocates to the watershed protection program. Otherwise the possibility increases that other goals usurp priority. In that case there is the danger that, even having implemented a watershed management program, the water supply is not adequately or most efficiently protected.

Nevertheless, urbanizing watersheds, almost by definition, do not exist solely to provide water; they are also part of other economic, social, and environmental communities. Thus, a water supply catchment area management program is likely to be implemented in the context of broader water resource management, environmental management, and comprehensive planning. Multiple and often conflicting objectives will exist. The problem is one of determining and attaining a pattern of uses of the watershed that is compatible with other objectives as well as the provision of good quality drinking water. Alternatively, of course, the implementation program could attempt to downplay or change other objectives.

Optimizing the use of the watershed for the collection of raw water for a potable water supply alone would generally preclude most agricultural and urban uses. Implementing such a solution, say through land acquisition and strict control of access to the watershed, would certainly not optimize other objectives and would be excessively costly besides. Feasibility would be extremely low. Consequently, net benefits, overall, would probably not be maximized by such a strategy. By contrast, it is conceivable that one could go to the other extreme and allow any and all uses of the watershed consistent with other objectives and then seek to meet the drinking water quality objective by treatment alone. That solution would probably be less than optimal as well. Treatment costs and the risk of undetected hazardous pollutants could be very high. Obviously, the optimal solution will be somewhere between the latter extreme of virtually no land management (accompanied by maximal water treatment) on the one hand and total prohibition of agricultural and urban uses (with minimal water treatment) on the other hand.

The water system may take any one of several strategies for acknowledging the reality of accommodating multiple uses of the watershed and multiple objectives. One way would be to delay expanding the goals beyond the water supply goal until the implementation planning stage. In selling the program and gaining the necessary allies to assure implementation, the goals would be expanded and management programs adjusted to reflect additional objectives. This approach retains a focus on the major goal through much of the program formulation process. It might, however, cause delays later, in the implementation stage, by requiring further time-consuming basic studies and reformulation of the management program. These delays could possibly be avoided by initial recognition of the multiple use-multiple objective nature of the problem, although requiring more time and resources in the early problem analysis and program design stages.

Another strategy would recognize the problem from the start, but would couch the other goals as constraints. In this way the water supply system's planning program would pursue the water supply goal while meeting minimum criteria to assuage other interests in the watershed. For example, a goal might be to protect raw water quality, while allowing reasonably productive agricultural uses of privately owned watershed land and recreational use of the publicly owned impoundment and its shoreland.

As mentioned earlier, it is important to obtain some consensus about the need to do something, even if in pursuit of different goals, from whatever governing bodies will eventually be expected to adopt and execute a watershed management system. In the absence of such a commitment, the danger is much greater that implementation will be unfeasible later. The need for agreement on the necessity for managing the watershed implies that the process of setting the focus and scope of those goals is not primarily a technical task. Rather it is a participatory process aimed at clarifying the values to be pursued in justifying the cost and effort of formulating and implementing a watershed protection program.

Establishing Solution Criteria to Supplement the
       Primary Goals

Regardless of the focus and scope of the community goals, a responsible water system manager and policy makers should add three fundamental criteria for formulating solutions:

1. Efficiency

2. Equity

3. Feasibility

These three criteria are very much "means-oriented." They emphasize that in addition to achieving the basic community goal of, say, an adequate supply of potable water, the watershed management program (the means) should be efficient, fair, and implementable.

The efficiency criterion forces explicit recognition
that the ends do not justify any and all means. This
criterion includes the concept of economic efficiency. The
net economic benefits that are obtained through a watershed
management system provide an important criterion. Cost-
effectiveness in achieving any minimum standards of raw water
quality is another useful criterion. The efficiency
criterion can be interpreted even more broadly, however, to
include a general economy of means. It suggests, for
example, that the simplest and least restrictive controls be
used, consistent with achieving the public purposes being
sought.

The equity criterion urges that the stream of costs and
benefits flowing from the management system be equitably distributed among property owners and non-owners, water system
users and non-users, and among the political jurisdictions
involved in the catchment area. It suggests, in one sense,
that windfalls and wipeouts be minimized and that those
enjoying the benefits also pay the costs. In another sense,
it suggests a sensitivity to the income redistribution goal
of society which argues that costs be distributed according
to ability to pay and the benefits according to need.
Equity, in the sense of fairness, also argues for procedures
to give fair advance notice to affected parties, an opportunity to state one's case, and routes for recourse if the
party feels unfairly treated by watershed controls.

The feasibility criterion is derived from the action-
oriented perspective of this Guidebook. It presumes that
water supply management alternatives should not only be
effective, economic, efficient, and equitable; they must also
be feasible. Feasibility refers to the possibilities for
adopting and effectively executing a management system.
Ultimately, desired ends must be balanced against efficient,
equitable, and feasible means.

Feasibility has several dimensions--financial, legal,
administrative, and political. Financial feasibility refers
to meeting the constraints on revenue raising capability of
the political jurisdictions which are financing the watershed
management sytem, on their tax bases and rates, and on their
unused bonding power. Legal feasibility refers to statutory
authority and constitutional limits to the use of governmental power. Administrative feasibility refers to staying
within the limits of the likely capability of staff and
decision-making boards to carry out monitoring and enforcement procedures that are implied in watershed control strategies and to design and execute public investments. Political
feasibility implies staying within the limits of achievable

consensus and commitment to goals and the limits of tolerance
to types of means. For example, tolerance of intervention in
private property rights in rural areas is usually quite
limited. The feasibility of intergovernmental coordination
is another consideration. In almost every case the water
system has jurisdiction over only a portion of the catchment
area Thus, since a watershed management system will of
necessity involve multiple governments, intergovernmental
coordination is vital.

Feasibility criteria should not be viewed as "hard and
fast" constraints. They represent factors that can be
changed or overcome. A solution that is not feasible today
may become feasible next year or five years hence as, say,
financial capability increases. Or, solutions may be made
feasible as the result of a conscious effort. For example,
the watershed management program may include as one element
the changing of state enabling legislation to make another
element, say the extension of land use controls in the watershed, feasible later.

Feasibility, factors affecting it, and approaches to
planning for increasing feasibility of implementation are
discussed in more detail in Chapter VI.

Goal Reduction

Having achieved some explicit agreement on the scope and
nature of the primary goals to be pursued by a watershed
management program and having added the consideration of
efficiency, equity, and feasibility of means, the water
system manager should take on the more technical task of
"goal reduction." This task involves deducing more explicit
and directly useful policies, criteria, and standards that
are consistent with the goals. Goal reduction is not
primarily a process of substituting or reassessing values,
although it might reveal misplaced values, just as the
evaluation of alternative solutions later on may shed further
light on the basic goals being sought.

The task of goal reduction begins with the determination
of objectives. An objective is a more rigorous, explicit,
measurable, achievable expression of a goal; it is an
intermediate milestone on the way toward a goal. A goal can
be divisible into a number of objectives. For example, the
goal of a safe yield might be expressed in the objective of
"x" million gallons per day by 1985 and "y" million gallons
per day by 1995. Water quality goals for the water supply
might be expressed as objectives by specifying the maximum

118

acceptable levels for specific primary and secondary contaminants, perhaps utilizing those specified at the federal or state level.

After interpreting goals in terms of specific objectives, the goal reduction process splits into two branches. One branch formulates design guidelines to aid in the search for alternative solution ideas and the other formulates criteria for evaluating alternatives and choosing among them.

Path One: Formulating Design Guidelines

On the first path, goals and objectives are interpreted as design principles and standards intended to guide the search toward more fertile regions of "solution space." Although they do not ensure optimum solutions, the principles are intended to facilitate the search for more promising solutions. The guides are derived from objectives; from efficiency, equity, and feasibility criteria; and from the problem analysis.

Perhaps it is best to begin with several "near universal" principles of water supply watershed management solutions:

1. In order to be effective, watershed management must cover the entire catchment area. Uncontrolled activities on one part of the catchment area, even a relatively small part, can negate the benefits obtained through costly and sustained efforts over a majority of the area.

2. The formulation of solutions should include ways to deal with the obstacles or constraints to action identified in the initial scoping step of the problem analysis or in the initial statement of goals, constraints, and solution criteria. Three basic categories of obstacles have been identified in Chapter II of this Guidebook: lack of technical information, political opposition to watershed management, and administrative shortcomings which hamper local action. The first of these categories of barriers is addressed in the problem analysis task, but the other two are aspects of the watershed management problem that directly affect feasibility and must be reflected in the form of technical solutions and implementation

119

strategies. In other words, watershed management proposals should address not just the technical water quality problem but also the problem of taking action—i.e., the institutional/political aspects of adopting and executing a course of action to protect drinking water supplies. This principle is explored in depth in Chapter VI.

The survey of local governments revealed that large proportions of respondents perceive serious obstacles to watershed management coming from opposition by major industries (73 percent), legal constraints in state law (65 percent), and opposition by agricultural interests (65 percent). The design of a watershed management program in any particular place must address whatever other critical deterrents to implementation exist there, including perhaps the large size of watersheds, lack of appropriate professional personnel, insufficient technical assistance from higher levels of government, and opposition from real estate, building, and land development interests, all of which were cited by over 40 percent of respondents in our survey.

The plan should provide a course of action either to (1) recognize and avoid the obstacles, or (2) devise ways to remove them. If the first, then the solution merely recognizes the obstacles as a constraint assumption of the problem definition. If the second, the solution proposes what actions are necessary to achieve sufficient change in the obstacle.

For example, a lack of jurisdiction over watershed territory might be overcome by a plan for cooperative agreement, creation of a new agency with adequate territorial jurisdiction and governmental powers, or other institutional mechanisms designed to remove the constraint.

3. The formulation of solutions must address all the dimensions of the problem. Thus, as the problem recognizes land uses, water resource needs, and institutional constraints/capacities, so must the proposed solutions.

4. The federal and state regional water programs, along with those of neighboring governments, are a part of the context within which a program must be designed and implemented. They represent both positive mandates and negative constraints, opportunities and resources as well as limitations.

The following examples taken from various sources illustrate the concept of design guidelines for particular places:

1. For protection from sedimentation, nitrogen, and phosphorous, for example:

   - Because over 50 percent of the anticipated nitrogen and phosphorous loading on the reservoir and 80 percent of the sediment is expected to come from irrigated cropland, the water management system should include measures aimed at agricultural practices.

2. Where on-site sewage disposal systems are a significant part of the water quality problem, for example:

   - The watershed management system should prevent new on-site disposal systems in areas where soil characteristics, land slope, or proximity to receiving waters will endanger their satisfactory operation.

   - The watershed management system should include measures such as monitoring and maintenance inspections, to ensure that performance standards for on-site disposal systems are being met.

3. To increase likelihood of attaining the feasibility criterion, for example:

   - Soil Conservation Service and Agricultural Extension agents should be involved in the design, adoption, and administration of any measures affecting agricultural practices. (Such a principle might be based on findings during problem analysis that farm community acceptance of control policies is better if traditional agricultural agencies are involved in the implementation of these measures.)

. The watershed management program should
employ measures that require the minimum
amount of change in current development
practices and have the lowest installation
and maintenance costs. (These principles
might also be considered as efficiency
promoting policies.)

4. To increase the likelihood of achieving the
efficiency criterion goal, for example:

. The program should employ the least costly
tools and interject the least interference
with the land market, consistent with
achieving a degree of control necessary to
serve the intended purpose. For example,
where clear health or safety objectives
exist, in the case of floodplains for
example, regulations are appropriate because
they are cheaper than public acquisition but
still achieve the purpose of protecting
people and property from an environmental
hazard and protecting water quality. Where
public access is required, say for
recreation, acquisition would be necessary;
regulations over the use of the land, while
cheaper, would not obtain public access.

5. To increase the likelihood of achieving the
equity criterion, for example:

. A program should cause as few "windfalls" and
"wipeouts" as possible, regardless of their
legality.

. Costs should, as much as possible, be distributed to those receiving the benefits
(except for those strategies pursuing income redistribution).

The guidelines illustrated in Table IV-1 were developed
by Blackman and Blaha[5] for those water supply watersheds of
the Southern Piedmont where active growth is just beginning
and where it is feasible to focus measures on the location,
design, and construction of new development. They also
suggest that local control programs provide closer review of
site plans and more frequent field inspection of construction
sites than required by state programs to ensure proper
implementation of sedimentation and erosion control mesures.
Fees from grading permits, they say, can be used to defray

122

Table IV-1. Principles to Guide the Design of Water
Supply Watershed Management Programs in the
Southern Piedmont

1. Locational principles to guide growth in order to lessen
    nonpoint source pollution loads from new development and
    thereby reduce the need for capital intensive stormwater
    detention basins or treatment facilities:

    a) Limit growth within and guide growth away from
       water supply watersheds to the extent possible.

    b) Control land use within the watershed by zoning and
       subdivision regulations, tax incentives, land
       acquisition, and other growth management tools.

    c) Limit municipal investments in infrastructure (water
       and sewer lines, highways, mass transit, industrial
       and commercial centers) within water supply
       watersheds which may encourage rapid development.

    d) Guide development into the upper parts of the
       watershed and maintain forests and natural vegetation
       in the lower parts.

    e) Encourage compact growth patterns. Cluster and
       Planned Unit Developments concentrate residential
       units and leave large areas of natural open space for
       stormwater infiltration or for cluster septic system
       drainage fields. Compact urban development patterns
       reduce construction area and vehicle traffic, thereby
       reducing pollutant generation. In a residential area
       of a given size and population, compact development
       will be less polluting than a conventional
       subdivision on both a per acre and a per capita
       basis.

    f) Locate land uses on appropriate soils. To minimize
       nonpoint pollution loadings from residential areas,
       low density development should be encouraged in areas
       characterized by soils with relatively high
       infiltration rates. Highly impervious land uses such
       as shopping centers and apartments should be
       encouraged to locate in areas with soils of
       relatively low infiltration rates.

    g) Establish minimum lot sizes based on soil
       characteristics for proper functioning of septic
       systems.

123

Table IV-1 (continued)

h) Regulate development to protect environmentally sensitive areas such as steep slopes, groundwater recharge areas, flood plains, and areas with highly erodible soils through zoning.

i) Require the preservation of natural vegetation along streams and lake margins by ordinances.

j) Prohibit all channelization of wetlands and streams.

2. Site design principles to minimize the percent of impervious surface and maximize infiltration of stormwater on-site:

a) Development on steep slopes (25 percent) should be avoided. Roads should follow the contours and the landscape should be graded to promote infiltration rather than the traditional design which sought rapid runoff of stormwater.

b) Sites with poor drainage or seasonally high water tables should be left as open space.

c) Limit clearing and grubbing of trees and shrubs.

d) The use of infiltration and detention/retention techniques encouraged to promote on-site disposal of stormwater and maintain predevelopment runoff rates and volumes.

e) Minimize the percentage of impervious land cover.

f) Maximize the use of natural drainage in designing development.

g) Redirect roof rain gutter outfalls away from impervious surfaces and toward lawns if soil permeability is adequate.

h) Encourage the use of grass swales.

3. Sedimentation and erosion control principles to minimize the impact of construction activities on receiving waters:

a) The development should be planned to fit the site with a minimum of clearing and grading.

Table IV-1 (continued)

b) The development should be phased so that only areas which are being actively developed are exposed.

c) Existing cover should be retained and protected wherever possible.

d) Critical areas such as highly erodible soils, steep slopes, streambanks, and drainageways should be identified and protected.

e) Stabilize and protect disturbed areas as soon as possible.

f) Keep stormwater runoff velocities low.

g) Protect disturbed areas from stormwater runoff.

h) Retain sediment within the site area.

Source: Adapted from Douglas Blackman and David W. Blaha, "Guidebook for Protecting the Quality of Surface Drinking Water Supplies in the Southern Piedmont," Master's Project submitted in partial fulfillment of the requirement for the Master of Environmental Management in the School of Forestry and Environmental Studies, Duke University, Durham, N.C., 1981, pp. 31-35.

the costs of such a program, thereby increasing its feasibility.

Once general principles are identified, standards should be developed to provide more specific statements of principles. For example, "steep slopes" may be defined as those over 15 percent. "Appropriate soils" may be defined by specific soil association categories or by performance criteria for percolation that should be met in a test of suitability for septic tank systems. Such specificity can be very helpful in designing both physical measures and governmental interventions for the watershed management program. Where the problem analysis and scientific knowledge justify specificity, one should incorporate standards into the design guidelines.

Some principles and standards may be interpretable on maps. Such maps identify the spatial pattern of "most suitable" areas as well as "environmentally sensitive" areas that are implied by the design principles and standards. The maps can be very suggestive for formulating the appropriate pattern of acceptable uses and practices in the catchment area and in designing the management program of intervention measures.

As is implied in the discussion above, the formulation of design principles is not solely an extension of goals and objectives. Design principles and standards represent interpretations of both goals (ends) and the context of problem conditions and structure. In addition, the "target" sources, target pollutants, target locations, and target hydrologic processes that will be discussed in the next section of this chapter can be thought of as inputs to the principles to guide design, or even as a form of principles themselves.

Path Two: Formulating Evaluation Accounts

The design guidelines discussed above are presumed to help formulate better alternative management systems. There is no guarantee that they will, however. Consequences may run counter to what is implied by a principle. Furthermore, of several alternatives based on the same design guidelines, one may be significantly better than another. Thus, there is a need for evaluation principles and standards as well as design principles and standards. Derived from the same goals and objectives as the design principles, evaluation principles and standards are oriented more directly to outcomes. They are meant to be applied to the consequences associated with a management program, i.e., to their outcomes, both projected before implementation and monitored as they occur after implementation. Design principles and standards by contrast are input-oriented; they are "suggestive" inputs to the watershed management system design process. Indirectly, of course, the evaluative criteria can also help guide the initial formulation of alternatives.

The evaluation criteria are like a set of accounts by which alternatives might be assessed. The accounts are derived from the goals and objectives, but are expressed in more measurable indicators. For further discussion of evaluation and evaluation criteria, see Chapter VII on "Evaluating Watershed Management Programs."

Goals and objectives, constraints, solution criteria,

design principles and standards, and evaluation accounts
provide one form of direction to the task of formulating a
watershed management program. A second form of direction
setting is the specification of targets at which to aim the
program --target pollutants, target sources of those pollutants, target locations within the water supply catchment
area, and target hydrologic process. These targets are what
the management program will try to influence.

SELECTION OF TARGET POLLUTANTS, TARGET ACTIVITY SOURCES, TARGET LOCATIONS, AND TARGET HYDROLOGIC PROCESSES

This step is essentially a synthesis and extension of
the problem analysis discussed in Chapter III. The emphasis
here is on choosing which of four target areas--pollutants,
pollutant sources, locations, and hydrologic pathways--will
be the targets of the watershed management system. Target
choices then serve to guide the formulation of intervention
strategies and specific measures.

The choice of targets is based on two major considera-
tions--the relative importance of the target in the water
supply problem (i.e., the consideration of effectiveness) and
the target's amenability to being controlled (i.e., the consideration of feasibility). Both considerations are important and where their implications conflict, the two will have
to be balanced. To some degree there can be a tradeoff among
the four target areas. For example, there can be a decision
to bypass control of the location of the source of the pollutant in order to concentrate instead on controlling hydrologic modification--intercepting and treating the pollutant in
transport before it reaches the impoundment or treating the
water within the impoundment but before it reaches the
drinking water intake. Or, to illustrate with another
possible strategy, one could downplay attention to sensitive
locations, and instead prohibit potential pollution sources
from the catchment area altogether; or vice versa. Again,
these choices will be based on the two considerations of
effectiveness and feasibility.

The feasibility criterion has three aspects to it. One
is the susceptibility of the pollutant, the source, and the
hydrologic process to structural measures. A second aspect
is the difficulty of inducing the necessary engineering
measure by nonstructural (intervention) measures, say by
regulations, preferential taxation, education, or subsidies.
A third aspect is the feasibility of getting the intervention
measure adopted and administered in an an appropriate form.
All three of these conditions must be met in order for a

target pollutant, source, location, or hydrological process to meet the feasibility consideration.

To illustrate the factors of feasibility and effectiveness, consider street surfaces in the watershed as a potential target of a management program. Street surfaces typically rank high as an important source of particulates, metals, and chemicals, particularly in commercial and industrial areas. As a target of the watershed management program, street surfaces might therefore meet the effectiveness consideration in a watershed with substantial acreage in built-up areas. Street surfaces rank relatively low, however, in terms of feasibility. Most sweeping equipment in general use (i.e., non-vacuum equipment) is low in efficiency, obstacles are posed by parked cars, and street sweeping is not generally done with sufficiently high frequency in areas other than business districts. Street sweeping using vacuum equipment, frequent scheduling, and removal of particulates from the watershed (rather than sweeping them into the stormsewer) is expensive, as is end-of-pipe treatment of stormwater, and therefore may fail to meet the implementability aspect of the feasibility consideration.

In selecting the targets of watershed management measures, then, the water system policy maker must keep in mind both the relative importance of the target and the feasibility of affecting it with watershed management measures.

Target Pollutants

The problem analysis discussed in Chapter III will have identified the types of pollutants that threaten the water supply and estimated the rates at which they are generated and delivered to the impoundment. The pollutants will be identified on the basis of their physical and chemical properties (e.g., microbiological, sediment, inorganic solutes, organic solutes, and radioactivity). But they may also be identified according to whether they are continuous or not, whether they are existing in the water, projected, or not present or projected but posing overhanging potential threat (i.e., not normally evident in measurements of water quality, but probable in the event of an accident or other triggering event).

The choice of a pollutant as a target should depend on its seriousness in terms of public health effects, economic effects of loss of storage capacity and water quality, and environmental effects on streams, ponds, and lakes. Seriousness will depend on the duration and frequency of the periods during which standards are exceeded and the number of persons, fauna and flora exposed and vulnerable to the pollutant. Choice of a target pollutant should also depend on the susceptibility of the pollutant to being mediated by a management program and by the feasibility of adopting and properly administering the necessary intervention measures.

Target Activity Sources

Sources that generate the threatening pollutants are also potential targets of the management system. Management strategy and specific measures will vary according to the pollution source being targeted. Sources can be classified in many ways; a few are presented and discussed here. One should adapt these approaches for specific situations, however, or invent more appropriate substitutes. Those discussed here have been inferred from the literature and our case studies.

As a first cut at classifying target sources, they may be identified as either existing uses and practices or as new (future) development. New development is often better treated as a separate or semi-separate problem from existing development. New urban development can be handled differently and probably more easily than existing development. This is due to the possibility for incorporating control measures into development design from the start and the possibility for establishing private responsibility for reduction of pollution emission or its treatment on-site rather than relying on public action to do so. Also, a non-degradation principle may be legally defensible, rather than having to provide a more rigorously demonstrated rationale for reduction of already existing water quality impacts, because ameliorative measures will likely be less costly and more easily passed along as part of the general cost of new development.

On the other hand, focusing on new development will obviously fail to solve an already existing water quality problem. In those cases where the existing water quality is a substantial part of the problem, it will be necessary to address the tougher challenge of reducing the generation of pollution from already operating uses or providing public facilities for ameliorating that pollution. In any case, the measures that are most suitable for new development--zoning, subdivision control, building codes, erosion and sedimentation controls, for example--are ineffective on existing development. A different type of measure is necessary--for example, improving regulations with regard to litter, animal waste control, dumping, outdoor storage of materials, use of pesticides and herbicides, and illegal discharges into streams and storm sewers.

Another approach is to classify target sources by type of land use. Different pollutants and degrees of threat are associated with different types of land use. The classification of land use types in Table IV-2 might be adjusted to fit the particular situation; not all of these land uses are likely targets for every water supply watershed management program. The classification used in a particular watershed should be responsive to the complexity of the specific water quality problem, the complexity of the anticipated management program, and the detail with which the problem analysis has been carried out with respect to land use activities. For example, a simple two-category system of agricultural use and urban use is a useful and frequently applied approach because the measures applicable for agricultural uses will be quite different from those appropriate for urban uses, and the target pollutants associated with each are likely to differ as well.

Blackman and Blaha provide another suggestive illustration of ways to organize the classification of target sources by dividing watershed uses into three land use "systems"-- urban, suburban, and rural.[6] The areas classified as urban systems include already developed land and tend to require measures designed to correct existing water quality problems. The suburban system consists of urbanizing land and tends to require preventive measures that can be incorporated into land development controls. The rural system consists of agricultural uses and tends to require a separate set of Best Management Practices for cropland, pastures, and forests.

Another approach to classifying sources is to identify specific sites of high-yield sources.[7] This approach is suggested by the fact that there is a great deal of variation in pollution levels that cannot be explained by variation in land uses, as discussed in Chapter I. Thus, perhaps it is better to identify and then control rigorously those high-yield sites and activities. Such targets might include dumping of liquid or solid waste on land surfaces, either on-site or off-site from the activities that generate such waste; major leaks in sanitary sewers; direct discharge of liquid waste to storm sewers and receiving waters; and discharges in excess of permits. Targeting site-specific high-yield sources suggests a management strategy that places emphasis on investigative activities, i.e., field sur-veillance backed up by authority to impose sanctions on site-

Table IV-2. Possible Target Sources Classified
by Land Use Type

Silviculture

Agricultural

    Pasture
    Cropland, nonirrigated
    Cropland, irrigated
    Intensive animal care (e.g., dairy farms, chicken
      farms, hog parlors, feedlots)

Construction (including clearing and grading)

Mining

Transportation—highways, railroads, trucking centers,
    gas stations

Urban Uses

    Residential and other low intensity development using
      on-site wastewater treatment
    Residential using public sewer
    Recreation
    Commercial—gas stations and car washes, laundromats,
      restaurants, other

Industrial (perhaps divided further into particular
    industrial sectors)

    Chemical users—e.g., petroleum, chemical companies
    Uses for which BOD of effluent is high, such as
      breweries, canneries, distillers, laundries, food
      and milk processing, pulp and paper, tanneries, or
      textiles

Solid Waste Disposal Sites

Sewage Treatment Plants

---

specific perpetrating activities. For example, instead of or in addition to classifying target sources such as industry or dairy farming and developing regulations aimed at all such uses, one would identify a specific food processing plant, let us say, or specific dairy, or a specific filling station that was dumping chemicals into the storm sewer or sanitary sewer, and then go after the offending establishment. The U.S. EPA stresses the importance of frequent sanitary surveys to identify and locate sources of health hazards in the watershed.[8]

Target sources of pollution problems can be couched along other dimensions as well, for example:

1. Population level

2. Population density

3. Employment level

4. Area of impervious surface

5. Point sources and nonpoint (or storm activated runoff) sources

6. Recorded and unrecorded pollution loadings (the latter often being subsumed under nonpoint sources for measurement purposes, but conceptually quite different)

7. Type of wastewater treatment from which pollutant originates (on-site vs. public sewerage)

Target Locations

In addition to target pollutants and target sources of those pollutants, it may be useful to specify target locations. One approach is to specify "sensitive areas," i.e., areas within the watershed that are sensitive by virtue of soil, slope, or proximity to the impoundment, the raw water intake within the impoundment, or feeder streams to the impoundment. Sensitive areas within the watershed are more vulnerable to certain activities in the sense that such an activity or practice in a "sensitive" location is likely to impact on water quality much more significantly than the same activity located at a different location within the watershed. Some sensitive areas, say those adjacent to the impoundment and feeder streams, may be vulnerable to a broad

132

range of activities. Other areas constitute an activity-
specific sensitivity, e.g., sensitivity to septic tanks due
to soil type or sensitivity to grading due to steepness of
slope. Of course, in a special way, the entire water supply
catchment can be thought of as a "sensitive" area compared to
other land in the region and therefore properly subjected to
special rules.

Ideally the specification of sensitive areas should
anticipate the needs of implementation devices. As such,
they should be defined in terms of features readily identifiable on the ground so that their boundaries can be clearly
identified in an ordinance or on a map. They should anticipate standards that can be used in legislation and display
features to which specific vulnerability can be convincingly
attributed (in order to justify to the community the
necessity to adopt the regulations and later to defend them
legally if necessary).[9]

Another use of the target location concept is the
mapping of land in the catchment area according to its
"suitability." Instead of specifying forbidden locations,
"suitability" suggests looking for the best locations for the
range of activities that are expected to be accommodated in
the watershed and then designing the land use management
system to encourage such activities to locate in those most
suitable areas and prohibit them in less desirable or "sensitive" areas. This is complementary to the designation of
vulnerable or sensitive areas from which certain activities
should either be excluded or required to undertake special
site design or engineering practices. In general it is much
easier to convert the sensitive area approach into management
measures than it is the suitability approach. Although both
are valid, the sensitive area approach is generally more
suited to water supply protection.

Target Hydrologic Processes

In addition to or instead of deciding on target pollutants, target sources, and target locations for the watershed
management system, some management systems may aim at a
fourth target. This fourth approach is aimed at con-
structively modifying hydrologic processes that transport and
transform pollutants, or conversely, preventing their
deleterious modification by urban development and agricultural practices. Hammer suggests that a helpful strategy for
areas of new development served by public sanitary sewerage
may even be to emphasize the hydrologic modification
strategy.[10] Such a strategy is based on the argument that

preventive measures to reduce and slow down runoff quantity will ordinarily prove to be effective in controlling water quality too, assuming that relatively high levels of control for public cleanliness (e.g., street cleaning and litter) can be maintained. Such a strategy would focus on control of erosion from construction sites; control of the location, design, and operation of on-site sewage disposal systems (particularly domestic systems); construction of leak-proof sanitary sewers; prevention of increases in peak discharge; prevention of increase in the base flow of streams; and protection of water courses from encroachment or modification (in part to maintain their capacity to serve as sinks for pollutants), all designed to maintain existing conditions as closely as possible.

The above approach assumes the need to maintain existing benign hydrologic processes. By contrast, purposeful intervention may also be designed to improve hydrologic processes, for example by aerating feeder streams or impoundments, treating water in retention and detention basins, or treating the water in the primary water supply impoundment. The South Fork Rivanna Reservoir serving the city of Charlottesville and Albemarle County, Virginia, for example, had an aeration system installed when the reservoir was constructed and later was advised to add copper sulfate to the reservoir to reduce effects of phosphorous loadings.[11]

FORMULATING AN APPROPRIATE COMBINATION OF USES AND PRACTICES FOR THE WATERSHED

Problem analysis has estimated the existing and projected conditions of concern and a causal structure among variables in the problem. Goals have been clarified and refined in the form of objectives and design principles, and supplemented by constraints and solution evaluation criteria. Target pollutants, target sources, target locations, and target hydrologic processes have been selected.

Let us assume that all of these tasks have been accomplished, at least in tentative form. The next consideration then is the formulation of an appropriate combination of uses, densities, development practices, and other practices that sufficiently alter the existing or policy-less projection of target pollutants, sources, and/or their locations, and hydrologic processes so that the watershed can be used efficiently as a water supply catchment area.

The design of appropriate uses and practices for the watershed is just a foot in the door, so to speak. In combi-

nation with the targets, goals, and principles, it comprises
a general policy or strategy. It does not, however, specify
the actual intervention measures—the actions to be taken by
one or more local governments to induce private and public
users and developers to undertake the desired combination of
uses and practices. The measures discussed in this section
are often called "physical measures" or "physical control
measures." The intervention techniques to be discussed in
the next chapter are then called "implementation incentives."*

There are four categories of physical controls:

1. <u>Specifying the land use pattern</u>—consisting of
   the types, locations, and intensities or
   densities of activities that should or should
   not occur in the watershed. For example, the
   watershed management program might specify that
   no multi-family residential development occur in
   the watershed.

2. <u>Specifying site design, site engineering or
   construction practices that should</u> occur—
   everywhere in the watershed or in sensitive
   areas; for all new development or for certain
   types of development everywhere or in sensitive
   areas only. For example, instead of prohibiting
   multi-family dwellings from the watershed, the
   program might specify site design standards,
   erosion and sedimentation controls, and storm
   water detention measures that, if provided,
   would render the multi-family development
   allowable.

3. <u>Specifying practices to be used in the on-going
   use of the land by agricultural or urban
   activities</u>. For example, the program might call
   for certain pavement cleaning practices in
   multi-family developments, inspection of
   detention basins and their operation, and
   monitoring of stormwater runoff.

4. <u>Specifying off-site "treatment" practices to be
   used in managing the feeder streams, water
   supply impoundments, and intake of raw water</u>.

Although a watershed management strategy may place

---

*For example, see reference 12.

significantly greater emphasis, or almost none, on one or another of these four approaches, the four categories are not mutually exclusive. They coexist as a mix within almost any management program and often apply in combination to the same target pollutant, source, location, or hydrologic process or to the same objective. For example, a program might allow industry only in certain locations, provided that certain site level engineering measures are installed, that the use or handling of certain chemicals in production processes is prohibited, and/or that treatment precautions are taken for impounded water.

Also, some specific intervention measures to be discussed in the next chapter may be legitimately classified in two categories. Zoning, for example, may specify the allowable types, locations, and densities of new development and therefore clearly be a category one control measure. Just as clearly, however, zoning often requires particular site design and engineering practices (category two) for certain uses or on certain sites. In other words, while the categories indicate distinctive strategies, some specific measures may fit in more than one strategy.

Physical Control Category One: Specifying the Type, Location, and Intensity of Activity and Development in the Watershed

This approach, which we call "locational planning" in the preface to the Guidebook, takes two forms. The first and simplest is the prohibition of some activities from the water supply watershed altogether. The second form divides the watershed area into compartments, each of which allows some activities and disallows others. The two forms are not contradictory. Thus, a management strategy may prohibit some activities and intensities entirely while allowing other activities or intensities in some locations but not in others.

The rationale for either the prohibition approach or the compartmentalization approach is to avoid inappropriate activities in inappropriate locations. In addition to water quality purposes, communities use locational control to foster development that can be efficiently provided with public services, avoids adverse impacts to surrounding properties, provides less costly development, and allows more efficient travel.

The prohibition approach should be considered for activities with especially high yield of a target pollutant and/or especially dangerous forms of pollutants. Activities and facilities that fall into this category include:

<u>Very high yield or potentially dangerous sources:</u>

Industrial activities that use known toxic, carcinogenic, or mutagenic substances

Commercial activities using similar materials

Land fills

Wastewater treatment plants

Feedlots and food processing plants

<u>Source activities that might be prohibited if counter-balancing forces for their inclusion in the watershed are not too great or if almost equally appropriate substitute sites outside the watershed are available:</u>

All industrial uses

Employment centers

Hospitals

Commercial centers—shopping centers, truck stops, clusters of gasoline stations

Warehousing and storage facilities

Heavily travelled highways

Dairy farms, hog parlors, chicken farms, veterinary facilities, and other intensive animal care facilities

High density housing

For those activities not prohibited from the watershed, there remains the task of designating where they should best be located. There may well be particularly sensitive locations from which an activity should be prohibited. There may also be other especially suitable places to which the activity ought to be steered. The device for establishing these determinations is a land use plan. Such a plan presents a land use pattern as a goal form for the future use of the watershed.

The land use plan may take on a range of precision; that is, it may range from a plan that suggests a few broad categories of land uses to a plan that suggests a greater number of districts or compartments in which the allowable uses are fewer and specifically named. Representing the style of few categories and least specific designation of activities is a "suitability/ critical area" plan. This type of plan designates critically sensitive areas in which a large number of uses, perhaps most urban and many agricultural uses, would be prohibited or carefully controlled. It also designates other so-called "suitable" areas which are most tolerant of many urban and agricultural activities. The designation of suitability may be made for particular urban or agricultural uses. For example, some areas may be designated suitable for agricultural activities but not urban uses. Other areas might be deemed suitable for urban uses served by a public sewer system but not suitable for development using septic tanks. Some areas might be suitable for a larger range of activities than are other areas.

Designation of restrictive areas--the critical environmental areas--should be based primarily on protecting the water supply. Designation of the suitability areas, on the other hand, can be based on a broader range of goals and objectives, but including water supply objectives.

A second and more definitive format for designating the desired pattern of activities is sometimes called a land classification plan. Such a plan makes choices from among several suitable locations for a particular activity, or to look at it differently, it selects an activity or small range of activities for a location from among a larger range of activities that are considered suitable.

The plan might classify land in the watershed into the following categories, for example:

1. Conservation areas or environmentally critical areas (a designation discussed already above).

2. Agricultural areas, sometimes divided between agricultural protection areas and less fertile areas where rural community centers and low density residential development might take place, utilizing on-site wastewater treatment.

3. Urbanizing areas, to which more intensive development, allowing or even requiring public sewerage, would be steered.

4. Built-up or already developed areas, to which further "in-fill" development would also be steered.

The intent in this approach is to divide the watershed into a few use-districts, thereby establishing a pattern of future uses that is compatible with maintaining the watershed's capacity to serve as a water supply catchment area.

The third and most detailed form of a land use plan divides the watershed territory into more compartments and makes more specific designations of allowable uses. The more detailed designations are normally concentrated on the built-up and urbanizing areas. These areas would be further divided into residential areas (of several densities perhaps), commercial areas, transportation facilities, perhaps industrial areas, and sites for community facilities such as parks and fire stations. The environmentally critical areas and agricultural-rural areas would remain as other non-urban designations in this full-fledged version of the plan.

Environmentally sensitive areas for urban water supply protection purposes would be part of all three forms of a land use plan for water supply watersheds. Such environmentally sensitive areas could include:

1. Impoundments and their shorelands

2. Feeder stream corridors

3. Particularly steep and otherwise more erodable areas

4. Impervious soils

5. Wetlands

Impoundments and their shorelands are, of course, commonly designated as critical areas from which all development and even active recreation uses such as fishing, motorboating, or swimming are often barred.

A 1978 proposal for the Delaware and Raritan Canal watershed provides an example of the use of the feeder stream corridors as an environmentally critical area.[13] (The D & R Canal serves as a water supply for one-fifth of New Jersey's population.) Variable width buffers are suggested to reflect the varying topographic and soil conditions along streams and wetlands. While small scale maps of the stream corridor

(1"=1000') would be available, exact determination of the
boundaries of the stream buffer area would be made in the
field, at the actual site, as part of the site analysis
process required for development approval. Figure IV-1 shows
a vertical cross-section through a typical stream corridor
(from an ordinance adopted by Middletown, N.J.); it
illustrates the factors by which the stream corridor width
and boundaries are determined for permanently flowing
streams. Less restrictive regulations hold for "headwater"
streams that do not flow throughout the year.

Rules and regulations are sometimes combined for feeder
stream corridors, impoundments, and impoundment shorelands.
They generally address cemeteries; storage, disposal, and
application of salts, herbicides, pesticides and other toxic
chemicals; human excreta and sewage treatment; storage and
disposal of radioactive materials; junkyards and refuse
disposal sites; bathing, swimming, boating and fishing; and
alteration of shores and channels.

Sometimes a more detailed water-and-land use plan is
addressed to uses and practices on impoundments and their
shorelands—a land use plan within a land use plan, so to
speak. Recreation opportunities often play a major role in
such a plan. Selection of appropriate activities for
impoundments and their shorelands must balance four factors:
requirements for water supply protection, the demand for
recreation and other opportunities that might be accommodated
on the impoundment and adjacent land, the suitability of the
critical area for various activities and developments, and
the scarcity of alternative locations not in water supply
watersheds. In protecting water quality, many factors must
be considered including whether the reservoir is a terminal
impoundment (containing the raw water intake) or a
collection-storage reservoir which feeds the terminal
reservoir; whether the demand is for land-based recreation
(and what kind) or water-based activities, and if so whether
water contact is involved; the probable adequacy and
reliability of housekeeping operations and waste treatment at
the recreation sites; as well as the water purification
system, and the size and conformation of the impoundment,
among other factors. Tables IV-3 and IV-4 provide an initial
guide to identifying recreation activities that are
compatible with the protection of the drinking water supply.*

---

*For further guidance on planning for recreation at
water supply reservoirs see references 14/15. For general
guidance in preparing a land use plan, see 16.

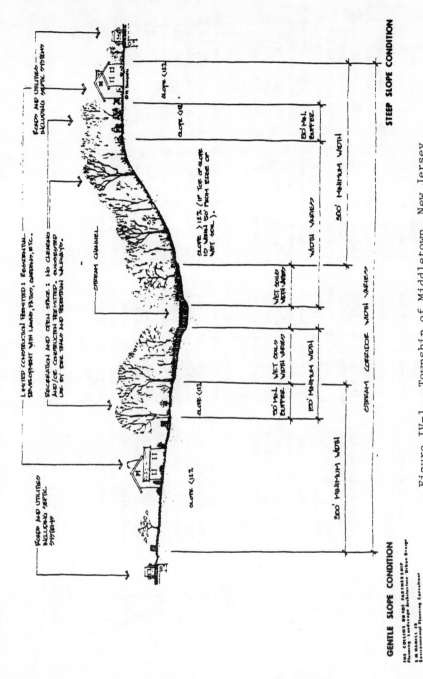

Figure IV-1. Township of Middletown, New Jersey
Mainstream--Stream Corridor

141

Table IV-3. Permissible Recreation at Collection Reservoirs

Reservoir Size	Plain Chlorination	Free Residual Chlorination	Filtration and Free Residual Chlorination	Activated Carbon or Other Advanced Process
Small <5 mi^2	Informal land-based recreation: hiking, picnicking, bicycling, golf, game sports, hunting, riding Informal winter land-based recreation: cross-country skiing, snowshoeing, ice skating (semiartificial rink), snowmobiling (no access to shore)	Fishing, boating (no motor), sailboating Informal land-based recreation: hiking, picnicking, bicycling, golf, game sports, hunting, riding Informal winter land-based recreation: cross-country skiing, snowshoeing, ice skating (semiartificial rink) snowmobiling (controlled access to shore)	Fishing, boating (small motor), sailboating Informal land-based recreation: hiking, picnicking, bicycling, camping, game sports, hunting, riding, golf Informal winter land-based recreation: cross-country skiing, snowshoeing, ice skating (semiartificial rink) snowmobiling	Fishing, boating (small motor), sailboating Informal land-based recreation: hiking, picnicking, bicycling, camping, game sports, hunting, riding, golf Informal winter land-based recreation: cross-country skiing, snowshoeing, ice skating (semiartificial rink) snowmobiling
Large >5 mi^2	Informal land-based recreation: hiking, picnicking, bicycling, golf, game sports, hunting, riding Informal land-based winter recreation: cross-country skiing, snowshoeing, ice skating (semiartificial rink) snowmobiling (no access to shore)	Fishing, boating (small motor), sailboating, swimming Informal land-based recreation: hiking, picnicking, bicycling, camping, game sports, hunting, riding, golf Informal land-based winter recreation: cross-country skiing, snowshoeing, ice skating (semiartificial rink) snowmobiling (no access to shore)	Fishing, boating (any size motor), sailboating, swimming Informal land-based recreation: hiking, picnicking, bicycling, camping, game sports, hunting, riding, golf Informal land-based winter recreation: cross-country skiing, snowshoeing, snowmobiling, ice skating (semiartificial rink)	Fishing, boating (any size motor), sailboating, swimming Informal land-based recreation: hiking, picnicking, bicycling, camping, game sports, hunting, riding, golf Informal land-based winter recreation: cross-country skiing, snowshoeing, snowmobiling, ice skating (semiartificial rink)

Source: See reference 14.

Table IV-4. Permissible Recreation at Terminal Reservoirs

Reservoir Size	Plain Chlorination	Free Residual Chlorination	Filtration and Free Residual Chlorination	Activated Carbon or Other Advanced Process
Small: Residence Time <30 Days			Fishing, boating (no motor), sailboating, swimming	Fishing, boating (no motor), sailboating, swimming
	Informal land-based recreation: hiking, picnicking, game sports, hunting, bicycling	Informal land-based recreation: hiking, picnicking, game sports, hunting, bicycling	Informal land-based recreation: hiking, picnicking, game sports, hunting, bicycling, golf	Informal land-based recreation: hiking, picnicking, game sports, hunting, bicycling, golf
	Informal land-based winter recreation: cross-country skiing, snowshoeing, ice skating (on reservoir), snowmobiling (no access to shore)	Informal land-based winter recreation: cross-country skiing, snowshoeing, ice skating (on reservoir), snowmobiling (controlled access to shore)	Informal land-based winter recreation: cross-country skiing, snowshoeing, ice skating, snowmobiling, ice fishing	Informal land-based winter recreation: cross-country skiing, showshoeing, ice skating, snowmobiling, ice fishing
Large: Residence Time >30 Days		Fishing (from shore)	Fishing, boating (no motor), sailboating, swimming	Fishing, boating (small motor), sailboating, swimming
	Informal land-based recreation: hiking, picnicking, team sports, hunting	Informal land-based recreation: hiking, picnicking, game sports, hunting, bicycling	Informal land-based recreation: hiking, picnicking, game sports, riding, hunting, bicycling, camping, golf	Informal land-based recreation: hiking, picnicking, game sports, riding, hunting, bicycling, camping, golf
	Informal land-based winter recreation: cross-country skiing, snowshoeing, ice skating (semiartificial rink), snowmobiling (no access to shore)	Informal land-based winter recreation: cross-country skiing, snowshoeing, ice skating (on reservoir), snowmobiling, ice fishing (controlled access to shore)	Informal land-based winter recreation: cross-country skiing, snowshoeing, ice skating (on reservoir), snowmobiling, ice fishing	Informal land-based winter recreation: cross-country skiing, snowshoeing, ice skating (on reservoir), snowmobiling, ice fishing

Source: See reference 14.

Physical Control Category Two: Specifying Site Design,
    Engineering and Construction Measures for New Land
    Development

The Category One approach of controlling the location of
activities does not eliminate the necessity to consider
preventive and mitigative measures at those sites when they
do become developed. Site level measures may therefore
supplement the locational strategy.

Site level measures involve the practices of grading and
clearing, construction, landscaping, paving, and the like on
the site, and the use of engineering structures and
landscaping. Such measures are intended to reduce the impact
of new development on hydrographic processes by reducing the
level of pollutants generated, reducing the pollutants
leaving the site either to affect water quality in the water
supply impoundment or to require the building of mitigation
structures by local governments off-site before the water
reaches the terminal impoundment; and decreasing the negative
modification of hydrologic processes. Site level measures
can be grouped according to three major categories and
several sub-categories:

1. Control of erosion, both during construction
    activity and thereafter, with properly designed
    grading, drainage ways, and landscaping.

2. Control of hydrologic modification by affecting
    storage, infiltration, and the general
    housekeeping at the site (to remove pollutants
    before they become activated by stormwater
    runoff). These measures generally do one of
    three things:

    a. Prevent an increase in peak discharge from
        what existed before development of the site
        and prevent a decrease in the base flow of
        streams and aquifer recharge through
        groundwater infiltration.

    b. Protect existing water courses from
        encroachment and alteration and stabilize
        channels and banks.

    c. Reduce the amount of pollution that will be
        activated by stormwater runoff.

3. Control of the location, design, construction,
    and even the operation of sewage disposal

systems (although operation will also be included in Category Three measures). This is generally done by:

a. Prevention of new on-site disposal systems in areas where soil characteristics, land slope or proximity to receiving waters will preclude satisfactory operation. (This strategy may already be reflected in the Category One: Locational Planning approach, but may be implemented through Category Two approaches as well. That is, one may wait to determine whether the site is suitable for on-site sewage disposal until development is proposed on specific sites and utilize the better information available at that time compared to what had been available earlier when the land use plan was prepared.)

b. Control over the design of new systems and appropriate supervision over construction in order to assure adequate performance under given land conditions and land use activities.

c. Maintenance of performance standards for on-site system operation (which may require monitoring and generally be included in Category Three measures below).

d. Promotion of recycling wastewater effluent and sludge.

Table IV-5 organizes specific site level measures under these categories. Space limitations and the Guidebook's emphasis on approach rather than detailed technical instructions preclude much discussion of the measures. References are provided in the table to sources of descriptions, discussions, and guidance on application of the measures.

Physical Control Category Three: Specifying Site-Level Practices for Existing and Future On-going Agricultural and Urban Activities

A third category of physical control deals with the site-level practices also, but unlike Category Two, these practices are aimed at on-going activities, not new development.

Table IV-5. Physical Control Measures to Protect
Water Supply Quality

Control Measures[a]	References to Descriptions, Discussions, and Guidance on Application

1. Erosion and Sedimentation Control

General good management of construction sites by minimizing disturbed areas and protecting trees and vegetation, appropriate phasing of grading and clearing, removing and stockpiling topsoil from areas to be paved, disturbed or affected by construction traffic, controlling construction traffic, road alignment and parking areas; providing crushed stone, temporary cover crops, or permanent landscaping for areas not actually in use for construction; ensuring slope stability; providing temporary runoff diversions and chutes; temporary sediment filters, trapping devices, and sediment basins.

Tourbier and Westmacott, 1981, pp. 69-82.

Finished grading: slope configuration and modifications, and runoff diversion measures.

Tourbier and Westmacott, 1981, pp. 83--86.

Vegetation and seeding measures for regraded areas: grass and grass legumes; hydroseeding and chemical stabilization, sodding, mulching.

Tourbier and Westmacott, 1981, pp. 87-98.

Protection of drainage channels and grass waterways: appropriate design, in-channel anti-erosion devices, check dams and protection of culvert mounts and chute outlets.

Tourbier and Westmacott, 1981, pp. 99-112.

2. Control of Hydrologic Processes

a. Measures to control increases in runoff and decreases in infil-tration:

Table IV-5 (continued)

Delays of runoff at source by temporary storage on roofs or in detention tanks, diversions, terracing, spreading, surface roughening, vegetative cover.	Tourbier and Westmacott, 1981, pp. 19-20, 29-30
Infiltration of runoff at source: removing or altering guttering and downspouts, dutch drains, rock-lined channels, porous asphalt, modular paving, grass swales.	Tourbier and Westmacott, 1981, pp. 21-28
Reduction of runoff and increase in infiltration after preliminary concentration: seepage beds, basins, pits, dry wells, areas and ditches; tile fields and perforated pipe; wells and gravity or pressurized shafts.	Tourbier and Westmacott, 1981, pp. 29-42
Delays in runoff after preliminary concentration: check dams, small up-stream impoundments, vegetated dry impoundments (detention basins), storage in sewer systems or in basins supplemental to sewer systems (applicable in combined sewers), parking lot storage, recharge of excess runoff by pressurized injection into aquifers.	Tourbier and Westmacott, 1981, pp. 43-60; Hammer, 1976, Sections 8 and 9
b. Measures to protect existing water courses from encroachment and alteration and to stabilize channels and banks:	Tourbier and Westmacott, 1981, pp. 113-128
Control of filling and development in floodways and floodplain, use of streams for dumping or direct watering of livestock.	Tourbier and Westmacott, 1981, pp. 161-166
Stabilization of stream channels and banks through vegetation, mechanical measures, and structural measures.	Tourbier and Westmacott, 1981, pp. 113-128.

Table IV-5 (continued)

c. Measures to reduce the amount of
    pollution in stormwater runoff:

Removal of pollution on-site: surface sanitation, and general litter control, including buoyant materials in the floodplain, disposal of leaves and garden refuse, garden chemicals, excreta from domestic pets.	Tourbier and Westmacott, 1981, pp. 131-132
Control of street and parking surfaces: motor vehicle contaminants, use of de-icing salts, road sweeping, isolation of spills, catch basin cleaning.	Tourbier and Westmacott, 1981, pp. 133-136
Treatment: removal of floating and suspended solids from runoff by use of trash racks, debris barriers, sediment ponds, screening, swirl concentrators, etc.	Tourbier and Westmacott, 1981, pp. 127-138
Reduction of and special practices for use of chemicals in industrial and agricultural activities, lawn care, highway de-icing, and service stations.	

3. Control the Location, Design, Con-
    struction, and Operation of Sewage
    Systems

Measures to conserve water, reduce sewage generation, and reuse gray water.	Tourbier and Westmacott, 1981, pp. 141-142
Prevention of on-site waste disposal systems where the locations are inappropriate due to soils, slopes, or proximity to receiving waters.	Hammer, 1976, Section 10
Use of waste disposal methods that do not use water.	Tourbier and Westmacott, 1981, pp. pp. 143-144

Table IV-5 (continued)

Proper design, installation, maintainance, and operation of on-site waste water disposal systems: anaerobic septic tanks, aerobic systems and other single family systems, elevated sand mounds, sand filter trench, evapotranspiration bed, stabilization ponds, multi-family cluster systems, etc.	Tourbier and Westmacott, 1981, pp. 139-168 Hammer, 1976, Section 10
Construction of leak-proof sanitary sewer systems.	
Catch basin and storm sewer design.	

aGeneral references for descriptions, discussions, assessments, appropriate target sources, target hydrologic processes, concerning on-site controls:

F.X. Browne, Water Quality Management Study of the South Rivanna Reservoir and Tributary Areas, Betz Environmental Engineers, Inc., Plymouth Meeting, Penn., June 1977.

Thomas R. Hammer, Planning Methodologies for Analysis of Land Use/Water Quality Relationships, Water Planning Branch, U.S. Environmental Agency, Washington, D.C., 1976.

Jochen Kuhner, Russell deLucia, and Michael Shapiro, "Assessment of Existing Methodologies for Evaluation and Control of Watershed Land Use in Drinking Water Supply Systems," in Robert B. Pojasek, ed., Drinking Water Quality Enhancement Through Source Protection, Ann Arbor Science Publishers, Inc., Ann Arbor, Mich., 1977.

J. Toby Tourbier and Richard Westmacott, Water Resources Protection Technology: A Handbook of Measures to Protect Water Resources in Land Development, ULI--the Urban Land Institute, Washington, D.C., 1981.

The approach applies to both urban and agricultural uses. There is a strong correlation, however, between the control of pollutants from agricultural uses and the Category Three approach. That is, almost all controls for agriculturally originated pollution can be put into Category Three, while Category One and Category Two approaches apply largely to urban uses.

Site level controls of on-going activities, like site level controls for new development, generally use "Best Management Practices" (BMPs). A BMP is a practice or combination of practices that is determined after problem assessment, examination of alternative practices, and appropriate public participation to be the best and most effective, practical (including technological, economic, and institutional considerations) means of preventing or reducing the amount of pollution generated by nonpoint sources to a level compatible with water quality goals.* BMPs may be structural (such as construction of artificial drainage ways or detention ponds) or nonstructural (such as no-tillage farming, contour farming and contour strip cropping). For agricultural uses, BMPs emphasize the control of sediment and pollution from surface runoff, generally resorting to soil and water conservation practices. For urban uses the targets are broadened to included the operation of on-site sewage disposal systems, general housekeeping of the land surface, the handling of toxic substances, the operation of site-level controls constructed during the development process, and dumping of liquids and solids on land or into storm sewers and streams. The discussion here will separate agricultural control practices from those aimed at urban sources.

With respect to formulating BMPs aimed at agricultural activities, the Soil Conservation Service (SCS) has identified several important factors to be considered:[17]

1. Adequate pollution control usually requires a system involving several conservation practices or measures.

2. There are usually several technically adequate alternative conservation systems that could physically be applied to any planning unit.

3. The best alternatives will depend upon such factors as:

   a. Land use
   b. Soil type characteristics
   c. Slope length and steepness

---

*Adapted from reference 18.

d. Landscape information
e. Land user's goals, financial assets,
    management abilities
f. Land ownership and leasing arrangements.

4. Development of best technical alternatives for
    a planning unit will usually require the
    knowledge and skills of a professional soil
    conservationist.

5. The land user should be included in the entire
    process of deciding which of several identified
    BMPs will be installed. The land user will
    apply and maintain the various components of
    the system; he will depend on the affected land
    for income; and he is the only person who can
    realistically make the system operate
    successfully over a long period of time.

In some states, BMPs are defined for use in large areas
of land that are characterized by particular patterns of
soil, climate, water resources, land use, and type of farming. Such areas are delineated by the SCS and classified as
Major Land Resource Areas (MLRAs). For example, five MLRAs
exist in North Carolina: (1) Blue Ridge; (2) Southern Piedmont; (3) Carolina-Georgia Sandhills; (4) Southern Coastal
Plain; and (5) Atlantic Coast Flatlands.

Any number of practices can be used to reduce water
pollution coming from agricultural uses in the watershed.
Table IV-6 provides a brief description of the conservation
practices that the SCS frequently utilizes to reduce nonpoint
pollution. Tables IV-7, IV-8, IV-9, and IV-10 describe the
level of effectiveness, suitability, and cost for the application of each of these specific conservation practices.

The ratings for effectiveness indicate relative levels
of efficiency. Ratings of suitability express the relative
need and suitability for each practice in different parts of
the particular state—North Carolina in this case. Average
costs for installing a BMP vary among geographic areas and
even within the same county because of differences in site
conditions and the availability of equipment and materials.
Annual costs for maintaining the potential BMPs are not
shown, but they will range from five to ten percent of the
original cost. Costs shown in the tables do not reflect
technical assistance (engineering services) for applying the
practices.

Table IV-6. Best Management Practices for Controlling
Nonpoint Pollution from Rural Land Uses

1. <u>Access Road</u> - A road constructed to minimize soil erosion while providing needed access.

2. <u>Chiseling and Subsoiling</u> - Loosening the soil to shatter restrictive layers and thereby improve water and root penetration.

3. <u>Conservation Cropping System</u> - Growing crops in combination with needed cultural and management measures to improve the soil and protect the soil during periods when erosion occurs.

4. <u>Conservation Tillage</u> - Tillage systems using some form of noninversion tillage that retains protective amounts of residue mulch on the surface throughout the year.

   a. <u>No-Till</u> -This system consists of planing in a narrow slot opened by a fluted coulter or other device in undistributed residues of the previous crop. Weeds are controlled with herbicides, resulting in a year-round protective cover of crop residues. No more than 10 percent of the soil surface is disturbed. The residues may be from meadow, winter cover crop, small grain, corn, or any other crop that supplies adequate amounts of residues for soil protection.

   b. <u>Minimum Tillage</u> - This method limits the number of tillage operations to those that are properly timed and essential to produce a crop and prevent soil loss.

5. <u>Contour Farming</u> - Farming sloping land in such a way that all operations are done on the contour in order to reduce erosion and control water.

6. <u>Contouring Orchard and Other Fruit Areas</u> - Planting orchards, vineyards, or small fruits so that all cultural operations can be done on the contour. The purpose of this practice is to reduce soil erosion, better control and use water, and to be able to use farm equipment more easily.

7. <u>Cover and Green Manure Crop</u> - A crop of close-growing grasses, legumes, or small grain used primarily for seasonal protection and soil improvement. It usually covers the land for a period of one year or less.

Table IV-6 (continued)

8. <u>Critical Area Planting</u> —Planting vegetation to stabilize the soil and reduce damage from sediment and runoff to downstream areas.

9. <u>Crop Residue Use</u> – Using plant residues to protect cultivated areas during critical erosion periods.

10. <u>Debris Basin (Sediment Pond)</u> – A barrier or dam constructed across a waterway or at other suitable locations to form a silt or sediment basin.

11. <u>Deferred Grazing</u> – Postponing grazing or resting grazing land for a prescribed period to improve hydrologic conditions and reduce soil loss.

12. <u>Diversion</u> – A channel with a supporting ridge on the lower side constructed across the slope to divert water and help control soil erosion and runoff.

13. <u>Fencing</u> – Enclosing an area of land with fencing to exclude or control livestock.

14. <u>Field Border (Filter Strip)</u> – A border or strip of permanent vegetation established as field edges to control soil erosion.

15. <u>Field Windbreak</u> – A strip or belt of trees established to reduce soil blowing.

16. <u>Grade Stabilization Structure</u> – A structure to stabilize the grade or control erosion in natural or artificial channels.

17. <u>Grassed Waterway or Outlet</u> – A natural or constructed waterway or outlet shaped and established in vegetation to safely dispose of water and runoff in order to prevent soil erosion.

18. <u>Heavy Use Area Protection</u> – Protecting heavily used areas by establishing vegetative cover, by surfacing, or by installing needed structures.

19. <u>Irrigation Water Management</u> —Determining and controlling the rate, amount, and timing of irrigation water, application to soil for plant needs in order to minimize soil erosion and control water quality and quantity.

20. <u>Livestock Exclusion</u> – Excluding livestock from an area to maintain soil and water resources.

Table IV-6 (continued)

21. Mulching – Applying plant residues or other suitable materials to the soil surface in order to reduce water runoff and soil erosion.

22. Pasture and Hayland Management – Proper treatment and use of pastureland or hayland to protect the soil and reduce water loss.

23. Pasture and Hayland Planting – Establishing forage plants to adjust land use, produce high quality forage and reduce erosion.

24. Planned Grazing Systems – A system in which two or more grazing units are alternately rested from grazing in a planned sequence to improve forage production and for watershed protection.

25. Pond – A water impoundment made by constructing a dam or by excavating a pit.

26. Proper Grazing Use – Grazing non-woodland areas at an intensity which will maintain enough vegetative cover to conserve soil and water resources.

27. Recreation Area Improvement – Establishing or thinning vegetation to improve an area for recreation use.

28. Recreation Trail and Walkway Protection –A pathway through a recreation area prepared for pedestrian, equestrian, and bicycle travel.

29. Spoilbank Spreading – Spreading excavated material over adjacent land.

30. Spring Development – Improving springs and seeps by excavating, clearing, capping, or providing collection and storage facilities.

31. Stream Channel Stabilization – Establishing structural work done to control aggradation or degradation in a stream channel.

32. Streambank Protection – Stabilizing and protecting banks of streams, lakes, estuaries, or excavated channels against scour and erosion by vegetative or structural means.

33. Stripcropping –Rowing crops in a systematic arrangement of strips or banks to reduce water and wind erosion.

Table IV-6 (continued)

    a.   Contour Stripcropping - Growing crops in a systematic arrangement of strips on the contour. The crops are arranged so that a strip of grass or small grain with legume is alternated with a strip of row crop. The purpose is to reduce erosion and runoff. Wildlife food and cover are also provided.

    b.   Field Stripcropping - Growing crops in a systematic arrangement of strips across the slope maintaining straight or nearly straight rows. The crops are arranged so that a strip of grass or small grain is alternated with a strip of row crop.

    c.   Wind Stripcropping - Growing wind-resisting crops in strips alternating with row crops or fallow and arranged at angles to offset adverse wind effects.

34. Terrace - An earth embankment, channel or a combination ridge and channel constructed across the slope to reduce erosion and sediment content in runoff water.

35. Tree Planting - Planting trees to conserve soil and moisture, or protect a watershed.

36. Trough or Tank - A trough or tank to provide drinking water for livestock at selected locations away from streams and ponds to bring about protection of vegetation and water resources.

Source: U.S. Soil Conservation Service, Potential Best Management Practices to Control Sediment Non-Point Source Pollution from Agricultural Land in North Carolina. Raleigh, N.C., 1978, pp. 8-11.

In spite of considerable experience with BMPs and the publication of numerous handbooks, there remains a lack of definitive information concerning the cost and long-term effectiveness of individual BMPs. This is particularly true for pollutants other than sediment, such as pesticides, nutrients, or heavy metals. In part, this situation reflects the difficulty of evaluating nonpoint source controls and of generalizing beyond specific case studies. Controlled studies on BMP effectiveness must contend with almost continuous change in the other factors affecting water quality, such as weather patterns and land use.

Table IV-7. Cropland

DEFINITION: Land used for all row crops, close-grown field crops, rotation hay and hayland, conservation use only, temporarily idle, orchards, vineyards, fruit, and openland formerly cropland.

Applicable Conservation Practices	Level of Effectiveness				Need and Suitability by MLRA as BMP					Cost Per Unit
					Rating of 1-3 With 1 Being Most Needed and Suitable					
	Sheet & Rill Erosion	Channel Erosion	Sediment Transport	Water Infiltration	MLRA 130 MTS	MLRA 136 SP	MLRA 137 SH	MLRA 133 SCP	MLRA 153 ACF	
1. Access Road	0	2	3	0	1	1	1	1	2	$0.20/LF
2. Chiseling & Subsoiling	2	0	0	1	2	2	1	1	1	$5/ac
3. Conservation Cropping System	1	3	2	1	1	1	1	1	2	$10/ac
4. Conservation Tillage										
a. No-tillage	1	2	2	1	1	1	1	1	2	$25/ac
b. Minimum tillage	2	3	2	2	1	1	1	1	2	$25/ac
5. Contour Orchard	1	3	3	2	1	1	1	3	3	0/ac
6. Contour Farming	1	3	2	2	1	1	1	2	3	0/ac
7. Cover & Green Manure Crop	2	3	2	1	1	1	1	1	1	$12/ac
8. Critical Area Planting	1	1	1	1	1	1	1	1	2	$900/ac
9. Crop Residue Use	2	3	2	2	1	1	1	1	1	$5/ac
10. Debris Basin (Sed. Pond)	3	2	1	3	1	1	1	1	2	$2500/basin
11. Diversion	1	2	2	3	1	1	3	2	3	$0.40/LF
12. Field Border (Filter Strip)	2	3	2	3	2	2	1	1	1	$0.12/LF
13. Field Windbreak (Wind Erosion)	1	0	1	0	3	3	1	1	2	$0.02/LF
14. Grade Stab. Structure	0	1	2	0	1	1	1	1	2	$1500/ea
15. Grassed Waterway	3	1	1	3	1	1	1	1	2	$800/ac
16. Land Use Change From Cropland to Grass or Woods	1	2	1	1	1	1	2	2	3	Grass- $70-100/ac Trees- $80-100/ac
17. Stream Channel Stabilization	0	1	3	0	1	1	1	1	2	$25/LF
18. Streambank Protection	0	1	3	0	1	1	1	1	2	$1.10/LF
19. Stripcropping										
a. Contour	1	2	2	1	1	1	1	3	3	$5/ac
b. Field	2	3	3	2	1	1	1	3	3	$5/ac
c. Wind	1	0	1	2	0	3	1	1	2	$5/ac
20. Terraces	1	2	2	3	2	1	3	1	3	$0.12/LF

Level of Effectiveness As Defined by EPA Region IV
1 - Most Effective
2 - Moderately Effective
3 - Least Effective
0 - No Effect

Need and Suitability by Major Land Resource Area (MLRA) As Best Management Practice (BMP)
1 - Suitable
2 - Moderately Suitable
3 - Least Suitable

Source: See reference 17.

156

Table IV-8.   Pasture and Hayland

DEFINITION: Land used primarily for production of hay or pasture from long-term forage stands.

Applicable Conservation Practices	Level of Effectiveness				Need and Suitability by MLRA as BMP					Cost Per Unit
					Rating of 1-3 With 1 Being Most Needed and Suitable					
	Sheet & Rill Erosion	Channel Erosion	Sediment Transport	Water In-filtration	MLRA 130 MTS	MLRA 136 SP	MLRA 137 SH	MLRA 133 SCP	MLRA 153 ACF	
1. Access Road	0	2	3	0	1	1	1	2	3	$0.20/LF
2. Critical Area Planting	1	1	1	1	1	1	2	1	2	$900/ac
3. Debris Basin	3	2	1	3	1	1	1	2	3	$2500/ea
4. Diversion	1	2	2	3	1	1	2	2	3	$0.40/LF
5. Fencing	2	1	2	3	1	1	2	1	2	$0.50/LF
6. Grade Stabilization Structure	0	1	2	0	1	1	1	1	2	$1500/ea
7. Grassed Waterway	3	1	1	3	2	2	2	2	3	$800/ac
8. Land Use Change From Pastureland to Woodland	1	3	2	2	1	1	1	2	3	$80-100/ac
9. Pasture & Hayland Planting	1	2	1	1	1	1	2	2	3	$70-100/ac
10. Pasture & Hayland Management	1	2	2	1	1	1	1	1	1	$40/ac
11. Pond	3	2	1	3	2	2	2	2	3	$3000/ac
12. Proper Grazing	2	2	2	2	1	1	2	3	3	0/ac
13. Spring Development	2	0	0	0	1	2	2	3	3	$150/ea
14. Stream Channel Stabilization	0	1	3	0	1	1	1	1	2	$25/LF
15. Streambank Protection	0	1	3	0	1	1	1	1	2	$1.10/LF
16. Trough or Tank	2	0	0	0	1	1	1	2	2	$200/ea

Level of Effectiveness As Defined by EPA Region IV
1 - Most Effective
2 - Moderately Effective
3 - Least Effective
0 - No Effect

Need and Suitability by Major Land Resource Area (MLRA) As Best Management Practice (BMP)
1 - Suitable
2 - Moderately Suitable
3 - Least Suitable

Source: See reference 17.

Table IV-9. Farmsteads

DEFINITION: Land used primarily for dwellings, farm buildings, lots, gardens, and other uses associated with farms and rural residences.

Applicable Conservation Practices	Level of Effectiveness				Need and Suitability by MLRA As BMP — Rating of 1-3 With 1 Being Most Needed and Suitable					Cost Per Unit
	Sheet & Rill Erosion	Channel Erosion	Sediment Transport	Water Infiltration	MLRA 130 MTS	MLRA 136 SP	MLRA 137 SH	MLRA 133 SCP	MLRA 153 ACF	
1. Access Road	0	2	3	0	1	1	1	2	2	$0.75/LF [1]
2. Critical Area Planting	1	1	1	1	1	1	2	1	2	$900/ac [2]
3. Diversion	1	2	2	3	1	1	3	2	3	$0.40/LF
4. Field Windbreak	3	0	1	0	3	3	1	1	2	$0.02/LF
5. Grade Stabilization Structure	0	1	2	0	1	1	1	1	2	$1500/ea
6. Grassed Waterway	3	1	1	3	2	2	2	2	3	$800/ac
7. Heavy Use Area Protection [3]	1	3	3	2	2	2	2	2	3	
a. Asphalt										$28,000/ac
b. Gravel										$8,400
c. Vegetation										$900/ac
8. Mulching	1	3	2	2	2	2	2	2	3	$200/ac
9. Ponds	3	2	1	3	2	2	3	2	3	$3000/ea
10. Recreation Area Improvement	2	3	2	2	3	3	3	3	3	$50-1000/ac
11. Subsurface Drains [4]	3	0	2	3	1	1	1	1	2	5"-$0.75/LF 6"-$0.85/LF 8"-$1.40/LF
12. Tree Planting	2	2	2	1	2	2	2	2	3	$80-100/ac
13. Temporary Debris Basin	3	3	1	3	1	1	1	2	3	$200-500/ea

Level of Effectiveness As Defined by EPA Region IV
1 - Most Effective
2 - Moderately Effective
3 - Least Effective
0 - No Effect

Need and Suitability by Major Land Resource Area (MLRA) As Best Management Practice (BMP)
1 - Suitable
2 - Moderately Suitable
3 - Least Suitable

[1] Based on all-weather gravel road, 14' bed. Includes grading and gravel.
[2] Areas such as roadbanks, gullies, critically eroding areas.
[3] Usually asphalt, gravel, or vegetation is used.
[4] Used for removal of surface water.

Source: See reference 17.

Table IV-10.   Other Lands

DEFINITION: All agricultural lands not included in cropland, pasture and hayland, farmsteads, woodland lands; includes wildlife land, recreation areas, and idle land.

Applicable Conservation Practices	Level of Effectiveness				Need and Suitability by MLRA as BMP					Cost Per Unit
					Rating of 1-3 With 1 Being Most Needed and Suitable					
	Sheet & Rill Erosion	Channel Erosion	Sediment Transport	Water Infiltration	MLRA 130 MTS	MLRA 136 SP	MLRA 137 SH	MLRA 133 SCP	MLRA 153 ACF	
1. Access Road	0	2	3	0	1	1	1	2	2	
a. Wildlife										$0.20/LF
b. Recreation Areas										Gravel - $0.75/LF
2. Critical Area Planting	1	1	1	1	1	1	1	1	2	$900/ac
3. Diversion	1	2	2	3	1	1	3	2	3	$.40/LF
4. Debris Basin (Sed. Pond)	3	2	1	3	1	1	1	1	2	$2500/ea
5. Field Border	2	3	2	3	2	2	1	1	1	$0.12/LF
6. Field Windbreak	0	0	1	0	3	3	1	1	2	$0.02/LF
7. Grassed Waterway	3	1	1	3	1	1	1	1	2	$800/ac
8. Grade Stab. Structure	0	1	2	0	1	1	1	1	2	$1500/ea
9. Heavy Use Area Protection	1	3	3	2	2	2	2	2	2	
a. Asphalt										$28,000/ac
b. Gravel										$8400/ac
c. Vegetation										$900/ac
10. Livestock Exclusion	2	3	3	2	2	2	2	2	2	$0.50/LF
11. Mulching	1	0	2	1	1	1	1	2	2	$200/ac
12. Pasture & Hayland Planting	1	3	2	1	1	1	2	3	3	$70-100/ac
13. Recreation Area Improvement	2	3	2	2	1	1	1	1	1	
a. Grasses & Legumes										$80-100/ac
b. Vines & Shrubs										Up to $1000/ac
c. Pruning Existing Trees										$50-100/ac
14. Recreation Trails & Walkways	2	0	3	3	1	1	1	1	2	$0.20/LF
15. Stream Channel Stabilization	0	1	3	0	1	1	1	1	2	$25/LF
16. Streambank Protection	0	1	3	0	1	1	1	1	2	$1.10/LF
17. Tree Planting	2	2	2	1	2	2	2	2	3	$80-100/ac

Level of Effectiveness As Defined by EPA Region IV
1 - Most Effective
2 - Moderately Effective
3 - Least Effective
0 - No Effect

Need and Suitability by Major Land Resource Area (MLRA) As Best Management Practice (BMP)
1 - Suitable
2 - Moderately Suitable
3 - Least Suitable

Source: See reference 17.

It is important to remember that design and selection of site-level controls should include many other factors besides the technical aspects of runoff and pollution reduction. Factors such as multiple use, maintenance requirements, safety, and aesthetics must also be considered in each case.

With respect to physical controls for on-going urban uses and urbanized areas, BMPs include the following:

1. General public and private property housekeeping practices to reduce litter along roads, vacant lots, unused parts of developed lots, and the like.

2. Paved street and parking lot cleaning practices.

3. Sewer and catch basin cleaning practices.

4. Maintenance practices for unpaved roads and ditches.

5. Reduction of dumping trash, garbage, chemicals, sewage effluent, and other pollutants onto watershed land, sewers, ditches, and streams.

6. Proper operation and maintenance of on-site sewage disposal systems.

7. Handling of hazardous materials.

8. Materials substitution in industrial processes.

Physical Control Category Four: Treatment of Pollutants by Public Agencies

To the degree that pollutants are still reaching receiving waters after Category One, Two, and Three controls have been induced, there remains the requirement for off-site treatment of the water. Such measures include:

Construction of off-site, in-stream detention structures designed to treat pollutants (at least allowing undissolved soils to settle out), as well as reduce peak volumes of stormwater runoff.

Introducing aeration or chemical treatment into
feeder streams or in impoundments before water
reaches the raw water intake. (For example,
the Charlottesville, Virginia, water supply
reservoir had an aeration system installed and
the Portland, Maine, Water District chlorinated
a feeder stream.)

Finished water treatment (see Chapter I).

SUMMARY

After problem analysis and before actually formulating
and implementing specific measures of intervention, a water
system should engage in:

1. Direction setting, in the form of setting goals
    and formulating management principles.

2. Target setting.

3. Specifying the appropriate uses and practices
    in the watershed.

We call this combination of activities formulating the policy
framework. It establishes the basic directions and form of
the watershed management strategy. It stops short of specifying the intervention measures by which the strategy is
given more definite shape, however. The focus is still on
clarifying ends (i.e., what needs to be accomplished),
deciding strategic targets to be influenced in order to
attain those ends, and specifying the combination of uses and
practices that would affect the target pollutants, sources,
and processes sufficiently to attain the purposes of the
management program.

Formulating a policy framework is an appropriate and
nearly indispensable step between analysis of the problem and
formulating specific governmental intervention measures. It
clarifies purposes and makes it easier to zero in on appropriate specific measures in the next stage. Without the
direction and basic dimensions of the program strategy
provided in the policy framework stage, there is the risk of
either too little focus to the search for specific measures
or of misplaced focus.

REFERENCES

1. Skidmore, Owings and Merrill. Assessment and Recommendations for Community Water Resources Planning, Office of Water Research and Technology, U.S. Department of the Interior, Washington, D.C., 1980, p. 8.

2. Guilford County Planning Board. Land Use Plan: A Strategy for Development in Guilford County, North Carolina, Guilford County Planning Board, Greensboro, N.C., 1966, p. 11.

3. "Reservoir Watershed Management Agreement" between the mayor and city council of Baltimore, Baltimore County, and Carroll County, Maryland, June 29, 1979.

4. Skidmore, Owings and Merrill. Assessment and Recommendations for Community Water Resources Planning, Office of Water Research and Technology, U.S. Department of the Interior, Washington, D.C., 1980, p. 126.

5. Blackman, Douglas and David W. Blaha. "Guidebook for Protecting the Quality of Surface Drinking Water Supplies in the Southern Piedmont," Master's project submitted in partial fulfillment of the requirements for the Master of Environmental Management in the School of Forestry and Environmental Studies, Duke University, Durham, N.C., 1981.

6. Blackman, Douglas and David W. Blaha, "Guidebook for Protecting the Quality of Surface Drinking Water Supplies in the Southern Piedmont," Master's project submitted in partial fulfillment of the requirements for the Master of Environmental Management in the School of Forestry and Environmental Studies, Duke University, Durham, N.C., 1981.

7. Hammer, Thomas R. Planning Methodologies for Analysis of Land Use/Water Quality Relationships, Water Planning Branch, U.S. Environmental Protection Agency, Washington, D.C., 1976.

8. U.S. Environmental Protection Agency. Manual for Evaluating Public Drinking Water Supplies, U.S. Environmental Protection Agency, Washington, D.C., 1975.

9. Kusler, Jon A. Regulating Sensitive Lands, Ballinger Publishing Co., Cambridge, Mass., 1980, p. 3.

10. Hammer, Thomas R. Planning Methodologies for Analysis of Land Use/Water Quality Relationships, Water Planning Branch, U.S. Environmental Protection Agency, Washington, D.C., 1976.

11. Miller, Todd L. and Raymond J. Burby with Edward J. Kaiser and David H. Moreau. Protecting Drinking Water Supplies Through Watershed Management: A Casebook for Devising Local Programs, Center for Urban and Regional Studies, The University of North Carolina at Chapel Hill, Chapel Hill., N.C., August 1981, pp. 305-306.

12. Bower, Blair and Daniel J. Basta. Analysis for Regional Residuals-Environmental Management: Analyzing Natural Systems, Resources for the Future, Washington, D.C., 1979.

13. Middlesex-Somerset-Mercer Regional Study Council, Inc.. Stream Corridor Protection in the Commission's Review Zone," A Memorandum to D & R Canal Commission, June 30, 1978.

14. Alford, Michael R. and Clark S. Binkley. "Planning Recreation at Water Supply Reservoirs," in Drinking Water Quality Enhancement Through Source Protection, Robert B. Pojasek, ed., Ann Arbor Science Publishers, Inc., Ann Arbor, Mich., 1977.

15. Newark Watershed Conservation and Development Corporation. The Pequannock Watershed Conservation and Development Plan, and Pequannock Watershed Conservation and Development Plan: Land Use Controls, City of Newark, Newark, N.J., 1975 and 1976.

16. Chapin, F. Stuart, Jr. and Edward J. Kaiser. Urban Land Use Planning, University of Illinois Press, Champaign-Urbana, Ill., 1979, especially Chapter 3, the introduction to Part III, and Chapters 11 and 12.

17. U.S. Soil Conservation Service. Potential Best Management Practices to Control Sediment Non-point Source Pollution from Agricultural Land in North Carolina, U.S. Soil Conservation Service, Raleigh, N.C., 1978.

18. Haith, Douglas A. and Raymond C. Loehr, eds. Effectiveness of Soil and Water Conservation Practices for Pollution Control, EPA-600/3-79-106, U.S. Environmental Protection Agency, Washington, D.C., October 1979, p. 1.

CHAPTER V

INTERVENTION MEASURES: THE HEART OF THE
WATERSHED MANAGEMENT PROGRAM

There are two major components in a water supply
watershed management program--decision guides and
intervention measures. Decision guides consist of:

1. Information on existing, emerging, and
    projected conditions in the watershed.

2. Goals, objectives, principles and standards.

3. Selected pollutants, source activities,
    locations, and hydrologic processes that will
    be particular targets of the management
    program.

4. Plans for the appropriate pattern of land using
    activities and practices.

5. A planned program of intervention measures.

All of these are intended to guide public and private
investment decisions and other practices in the watershed.
Perhaps the most important purpose of these decision guides,
however, is to guide the selection and implementation of
watershed intervention measures.

Intervention measures are the actions necessary to
complete the management program in fact. They exert direct
governmental influence on the course of events in the
watershed, especially on the way that land and waterways are
used. They consist of regulations, capital improvements,
pricing mechanisms, taxation schemes, bonus and penalty
provisions in regulations, education programs and
surveillance activities.

Decision guides are important, but intervention measures
are at the heart of the management program. They are the
action-forcing measures that coerce or induce public and
private watershed users to undertake the appropriate physical
controls discussed in the previous chapter.

165

The purpose of this chapter is to provide a systematic review of intervention measures that are appropriate for watershed management. The review is organized according to the major powers of local government: the police power (including development regulations and health regulations), compensatory powers (including acquisition of land), provision of community facilities (including extension of sewer lines and roads), and taxation (including preferential taxation for uses compatible with water supply protection). Under each major power, measures are described briefly with particular reference to their use for water supply protection. The discussion of both broad governmental powers and particular measures is designed to provide sufficient information so that the user of the Guidebook can decide whether, for his or her particular situation, the device is sufficiently relevant to warrent further exploration.

The review of intervention measures is not comprehensive. It does not include all measures that might possibly be used. Instead, the chapter concentrates on those tools and techniques that, in the judgment of the authors, will be most useful to water system managers and local planners. Thus, for example, compensatory regulations have been proposed for growth management and environmental protection and might be relevant to water supply protection, but they are not reviewed here because they are not readily useable within the political context and under the statutory authority that exists in most water supply watersheds.

Although the emphasis is on local government, some references to state and federal regulations and incentives are also included, since these can be important. In some cases they provide constraints on local actions. In other cases they may enable implementation of a particular measure or augment its effectiveness.

Table V-1 provides a listing of the powers and measures to be discussed in the remainder of this chapter.

REGULATORY MEASURES: APPLICATIONS OF THE POLICE POWER

Regulatory measures are the backbone of almost any land use management program, including most watershed protection programs. Regulations, properly administered and enforced, provide a certainty not available though incentive devices and hortatory approaches; they are generally less costly to local government than public capital investment programs, including water treatment plants; and their legal defensibility for water supply watersheds can be established

166

Table V-1. Intervention Measures

Regulatory Measures: Application of the Police Power

  Interim regulation (moratoriums)
  Zoning techniques

    Prohibition of industrial, commercial, or high
      density residential uses from watershed
    Large lot residential zoning
    Exclusive agricultural (nonresidential) zones
    Cluster zoning and planned unit development (PUDs)
    Conditional zoning and special use permits
    Special districts (sensitive area)
    Performance zoning

  Subdivision regulations
  Erosion, sedimentation, and stormwater controls
  Regulation of septic tanks and other on-site sewage
    disposal systems
  Regulation of solid waste disposal sites
  Regulation of wastewater discharges
  Regulation of impoundment surface and shoreline

Acquisition of Property Rights: Application of Compensatory
  Powers

  Fee simple acquisition
  Easements (less than fee simple acquisition)
  Transfer of development rights

Public Investment in Capital Improvements:
  Provision/Withholding

Preferential Taxation

Other Components of a Watershed Management Program

  Land surface sanitation practices of local government:
    street cleaning, sewer cleaning, and general clean-up
    operations
  Centralized management of the operation and maintenance
    of on-site disposal systems
  Encouraging voluntary land management practices by
    farmers
  Public relations, education, and participation
  Legal action

because regulations are well suited to "prevention of harm" and there is a clear relation between harm to drinking water quality and public health.

On the other hand, there are difficulties. First, the local government must have the statutory authority to use the particular regulation and to use it specifically for protecting the water supply. Since all local regulatory powers are derived from the police power of the state, a regulatory power may not be exercised by sub-state bodies until the state, through enabling legislation, has described the nature of the power and authorized its use by the local body. Water supply systems, for example, often do not have authority to impose land use regulations, unless they are also a part of a general purpose municipality, county, or a special district to which the state legislature has granted such authority. Even some general purpose governments, particularly the more rural counties in whose jurisdictions water supply catchment areas are often located, may not have such authority or if authorized may have little or no experience with land use regulations and may be disinclined politically to use them.

Second, regulatory power is constrained under the U.S. Constitution from being confiscatory. That is, the regulation must not so limit the use of land as to amount to a taking of private property for public use without compensation. While a dimunition in the value of the land is permissible, the ordinance, if challenged, will be held invalid by the courts if the regulation has the effect of completely depriving the landowner of the beneficial use of his property by precluding all practical uses or if the restriction is more burdensome on the property owner than is justified by the harm prevented.

A third possible difficulty with regulations is that they must meet the due process requirements of the Constitution. Due process requires that the exercise of governmental powers serves a legitimate purpose of protecting the public from health, safety, and moral hazards or protecting the general welfare. It also requires that the regulation be reasonably related to the objective and not be unduly oppressive. Further, safeguards and rules must be incorporated in the regulations with respect to public notice, opportunity for interested parties to be heard, and other procedural guidelines often set forth in the state's enabling legislation.

Fourth, in order to be consistent with the equal protection clause of the U.S. Constitution, regulations must not discriminate among landowners who are similarly situated.

Regulations may, of course, treat different pieces of property differently, but these differences must have some rational justification which relates back to the purpose of the regulation. They cannot be arbitrary. More on the legal and other limitations to the implementation of regulatory powers is provided in Chapter VI.

Interim Regulations

Interim regulations can play a limited but potentially important role in a watershed management program. Sometimes called moratoriums, they are used to temporarily slow or stop development in the watershed or part of the watershed until a planning process has been completed and a scheme of permanent controls has been devised and implemented. They serve several purposes. The first is to provide "breathing space" to allow staff to undertake the technical planning process and to learn from desirable public debate about watershed management rather than spending a good part of its time processing development regulations. The second is to prevent development that will be contrary to the eventual watershed management program from taking place before the program becomes operational. Without interim regulations, a proposed watershed management program may actually stimulate a rush of development proposals attempting "to get in before the gate closes." A possible third purpose would be to allow time to deal with a crisis situation, e.g., to prevent degradation of water quality that already fails to meet federal or state standards and that poses a clear and present danger to public health.

Interim regulations may stop development altogether in the entire watershed or they may be applied to critical areas or critical land uses. They could be applied to one of several critical points in the development process--rezoning requests, construction permits, subdivision plats, special use permits, septic tank permits, water or sewer hook-ups or building permits, for example. Such "temporary" restrictions on development approvals, pending the adoption of a permanent watershed management program, constitute a legally defensible exercise of the police power and are generally presumed by the courts to be a valid exercise of legislative prerogative. Interim regulations are particularly appropriate when there is a clear and present environmental problem, which may well be the case in the water supply situation. Interim regulations must be reasonable and sensibly related to the purposes being served, and the time limit must be reasonable, but may range from several months to two years or longer. Even if struck down, interim regulations may serve their purpose of

169

restricting development temporarily while a more comprehensive watershed management program is being designed and implemented.

Interim regulations have been in common use for over a decade. Often, however, they have been seen as a quick and easy solution, whereas they are not. They do nothing more than buy time and should not be conceived of as substitutes for a more carefully considered and permanent watershed management program. But the time they buy can be a critical contribution. Among the case studies on which this Guidebook is based, Albemarle County, Virginia, Baltimore County, Maryland, and West Milford (in Newark, New Jersey's watershed) successfully employed moratoriums.[1]

Zoning Techniques

Conventional zoning is one of the most traditional and commonly used land use controls in local government, along with subdivision regulations. Zoning divides a political jurisdiction into districts or zones, each of which places different restrictions on the type of land use allowed there, the density of development (by specifying maximum number of dwelling units per acre or minimum lot size), height and bulk of structures and their setback from the street and side and rear lot lines, minimum number of parking spaces, and other requirements. In this way zoning seeks to coordinate private and public development (e.g., making sure that streets and sewers will be adequate without having to oversize them to be ready for any and all development possibilities); to avoid undesirable side effects of development by separating incompatible uses, grouping compatible uses, and maintaining adequate standards for individual uses; and, most important, to assign land uses to sites having the most suitable environmental characteristics for those uses so that adverse environmental effects will be minimized. There are a number of specific techniques under the zoning umbrella and our discussion will henceforth be directed at these individual techniques rather than the general concept of zoning.

Industrial, Commercial, and High-Density Residential Zoning

In order for a zoning ordinance to be legally defensible, a locality must allow all or most legitimate activity somewhere in its jurisdiction. That need not be in the water supply watershed, however. Provided reasonably good alternative sites are available in the locality, a zoning ordinance can be used to prohibit industrial, commercial, and/or high-density residential uses in the watershed, allowing such uses in zones outside the watershed.

Large Lot Residential Zoning

Most state zoning enabling acts allow localities to regulate the size of lots in order to minimize adverse environmental impacts; to provide adequately for sewage disposal, water, schools, parks, and transportation; to prevent overcrowding of the land; to lessen congestion; and generally to promote health and general welfare. Thus, since the end of World War II the use of large lot zoning has become commonplace, perhaps even overused as a knee-jerk response to development pressure.

One of two purposes generally underlies the use of large lot zoning. In one it is used temporarily to defer development, establishing what is in effect a holding zone. The intent is to rezone the land eventually to allow more intensive uses as the community is able economically to provide water, sewer, and other facilities. The second use of large lot zoning envisions low density development as the ultimate use of the land so zoned. This second purpose is more relevant for water supply protection.

The minimum size of lots should depend on the sensitivity of the land to disturbance because of its soils, slopes, and proximity to feeder streams and the area required for adequate on-site disposal systems, including enough land to provide contingency sites for replacement systems. Almost all the case study watersheds were protected by some form of large lot zoning. Generally the minimum lot sizes were from one to three acres, based on the amount of land required for septic tanks to function properly and to provide new septic fields should the initial systems fail. Marin County, California utilizes ten-, twenty-, and sixty-acre lot sizes to discourage residential development in the Nicasio Watershed altogether and to preserve agricultural land.[2]

Large lot zoning is a familiar tool and one of the more easily implemented devices to keep down densities and thereby keep down the population in the watershed. Its use is based on the assumption that low density housing is less polluting than industrial, commercial, and denser residential development. However, other techniques that allow the clustering of housing onto the most suitable land can be more effective than large lot zoning for protecting the environment, providing that appropriate on-site engineering is employed. Clustering also creates open space systems and maintains more of a rural ambience.

Exclusive Agricultural (nonresidential) Zones

This type of zone excludes even low density housing and constitutes a direct control of the population level in the watershed. The justification would have to be based not only on the need to protect water supply, but also on unique suitability of the land to support agricultural uses consistent with protecting water quality. Feasibility is questionable in many places, but agricultural zoning has been used in Florida, California, and in northern Illinois in areas where agricultural productivity and value are high. In most areas of the country it is questionable whether the designation would stand up permanently under development pressures in an urbanizing watershed. Furthermore, maintaining a watershed or large parts of it in agricultural uses does not assure high quality raw water. Agricultural zoning should therefore be accompanied by other measures that induce appropriate agricultural management practices.

Cluster Zoning and Planned Unit Development (PUD)

This technique can be an alternative to large lot zoning and agricultural zoning or a supplement to them. Sometimes called average density zoning, it permits more flexible design of developments built as a unit. Development can be concentrated on the best parts of a site, thereby preserving the more environmentally vulnerable parts, as long as the total number of units does not exceed the number allowed in the ordinance. Density is transferred, so to speak, from more vulnerable areas of the site to the less vulnerable areas. Specific plans for the development are required in advance and must be approved by the permitting officer or body.

Clustered watershed development may be less polluting than low density residential land uses on large lots. Less land is disturbed when development is concentrated. Pollution controls, such as stormwater control devices and public sewerage, are also more feasible with dense development. However, water quality problems caused by poorly maintained stormwater and erosion control devices, and the threat that the level of urban development may continue unchecked, are countervailing arguments in favor of large lot zoning. In Baltimore County and Newark, court suits challenging the legality of large lot zoning are examining the water quality benefits afforded by large lot zoning compared to clustered development.[3]

Clustering provisions can be joined with agricultural

zones. An agricultural zone, for example, might allow one
dwelling per five, ten, twenty-five, or up to one hundred
acres, but allow the dwellings to be clustered on lots of one
or two acres on a portion of a much larger agricultural
parcel. This approach is used in the water supply watersheds
of Baltimore County, Maryland, Marin County, California, and
Albemarle County, Virginia. Farmers are allowed to subdivide
a portion of their land, the number of lots being determined
by a sliding scale and the size of their overall holdings.
The remainder of the holdings is then restricted to agricultural use.[4]

Conditional Use Zoning and Special Use Permits

These techniques allow approval of development to be
conditional, based upon provision of site-level design and
engineering features that protect water quality. It is useful when a particular use may not be objectionable, provided
certain precautions are taken to control runoff, assure
adequate waste handling, avoid more sensitive areas of the
site, or otherwise protect the water supply.

The types of uses that might be subject to conditional
use provisions in a water supply watershed are those having
high yields of potential pollutants and activities involving
the use of hazardous materials. Conditional permits for such
uses might govern operating procedures, such as parking lot
sweeping or operation of treatment facilities, with regular
renewal reviews required to maintain the permit. Examples of
potential uses to which this approach might apply include
certain agricultural uses that involve intensive animal care
(such as dairy farming, hog farming, and feedlots), gas
stations, petroleum distributors, mining, certain manufacturing operations, and certain recreational uses of the feeder
streams and raw water impoundments.

Special Districts

Special districts are sensitive areas in which most
types of development are prohibited, or allowed only upon
certain precautions being taken to limit density, provide
site design features, or construct engineering solutions to
eliminate the threat of environmental damage. In some cases,
the watershed management program might want to combine
special districts with conditional use permits so that in
some districts a particular use might be allowed by right
while in other special districts it is allowed only as a conditional use, i.e., only if acceptable precautions are taken.

173

Special districts that might apply to water supply protection include water supply watersheds per se, recreation districts, shoreland districts, stream corridor districts, districts with special restrictions on septic tanks due to soil conditions or proximity to water courses and impoundments, districts with special restrictions on dairy farming, hog farming, and other agricultural uses involving intensive animal care.

Performance Zoning

This approach sets standards for acceptable levels of side effects, such as the quantity, velocity, and pollutant level of stormwater runoff, rather than specifying the acceptable uses for the site or the specific design measures that must be taken. Attention is thereby focused on acceptable levels of environmental impact, perhaps zero impact, while allowing the market and the developer to determine the appropriate use and the appropriate site measures to meet impact standards.

Such an approach can be used in conjunction with some of the techniques already described above (e.g., the special use, planned development, and conditional zoning approaches) or as an independent approach to zoning of the watershed. Performance zoning would set standards for each zone rather than setting the list of allowable uses. Providing the prescribed standards are met, a wide range of uses could be permitted. For a water supply watershed, performance criteria might address the control of hazardous substances, stormwater runoff, and on-site waste disposal.[5/6]

Performance standards, while permitting more flexibility, are more difficult to administer because they require those who review and approve development and those who inspect and enforce development standards to have a good technical grasp of environmental processes and the wide range of runoff control techniques, waste treatment techniques, and the like, and to have the ability to exercise good judgment in determining whether a measure is best for a particular piece of land. A common approach is to place the burden of designing the measures and of proving their adequacy on the developer; however, even that approach requires a sophisticated capacity to review his proposals for adequacy.

Subdivision Regulations

Subdivision regulations control the division of raw land into buildable sites. Thus they are a way of controlling new development. Their primary purpose in watershed protection would be to stipulate design and engineering standards and even construction practices that must be met in order to obtain approval of a subdivision plat and which in turn would be required before buildable lots could be sold and recorded.

More specifically, subdivision regulations may be used to enforce minimum lot sizes and configurations required in the zoning ordinance; assure adequate stormwater runoff and drainage control through requirements for drainage easements, proper grading, on-site retention or detention of runoff, and improved drainage ways; and obtain minimum street right-of-way drainage specifications and other physical measures for stormwater control that were discussed in Chapter IV. Sub-division regulations and review procedures may be the vehicle for obtaining adequate design and installation of septic tank systems or other on-site wastewater controls; erosion and sedimentation control provisions; meeting environmental impact statement requirements; and meeting requirements for adequate public facilities. In this last regard, subdivision plat approval may be contingent upon the provision of adequate off-site as well as on-site facilities, including sewers, roads, water lines, parks, and schools. This approach is discussed further below under public investment in capital improvements. Subdivision requirements dealing with water quality objectives can be expressed either in terms of specific control measures that must be employed or as environmental performance standards.

Subdivision regulations can also require mandatory dedi-cation of land for neighborhood recreation. It is possible to utilize this feature to encourage dedication of stream corridors to provide buffers of natural vegetation between new development and streams feeding the water supply impound-ment.

Subdivision regulations, along with zoning, are the most commonly used local control on new land development. This constitutes an advantage to their use as a vehicle for appropriate site planning because they are already familiar to most elected and appointed local officials as well as those in the development industry and even landowners.

Erosion, Sedimentation, and Stormwater Controls

Most states have sedimentation and erosion controls that apply to new development, but this should not suggest that local attention is not needed, particularly within water supply watersheds. Local control programs should provide closer review of site plans and more frequent field inspection than state programs which are often not rigorously enforced. Guidelines to follow in formulating local ordinances for water supply watersheds include:[7]

1. The development should be planned to fit the site with a minimum of clearing and grading.

2. The development should be phased so that only areas which are being actively developed are exposed.

3. Existing cover should be retained and protected where possible.

4. Critical areas such as highly erodible soils, steep slopes, stream banks, and drainageways should be identified and protected.

5. Stabilize and protect disturbed areas as soon as possible.

6. Keep stormwater runoff velocities low.

7. Protect disturbed areas from stormwater runoff.

8. Retain sediment within the site area.

Stormwater and erosion control regulations were commonly used in site level pollution control programs in the case study communities investigated for this book. In highly urbanized areas such programs normally concentrated on controlling runoff from streets, parking lots, and other impervious surfaces and minimizing erosion from construction activities. Physical measures required by the regulations in the case studies included detention basins, grass swales, and storage of parking lot stormwater runoff. A broader range of physical approaches is discussed in Chapter IV.

Both design standards and performance standards were used in sedimentation and erosion control regulations in the case study communities. The design standards that form the basis for the Delaware and Raritan Canal watershed management program require specific types and designs of detention basins and structures.[8] The requirements vary by type of development and susceptibility of the site to environmental impacts. For example, larger commercial development must have detention basins capable of handling flows from a 100-year storm. In Albemarle County and in Baltimore County, performance standards stipulate an allowable level of pollution that can be generated but do not state how these

standards must be met; Baltimore County uses a mixture of
design and performance standards. Among the latter is a
requirement that the characteristics of runoff after
construction not be substantially different from its
characteristics before development took place.[9]

Regulation of Septic Tank and Other On-site Sewage Disposal Systems

On-lot disposal systems are prevalent in urbanizing
areas as well as in rural areas. The conventional septic
tank system and modifications of it is the most common
on-site system, but alternatives include the mound system,
evapotranspiration bed, low pressure pipe system, filter
system, cinder block system, or an aerobic system.[10]
Regulations should encompass any alternatives to septic tank
systems that are allowed.

Regulation should cover six considerations: (1) site
evaluation to determine suitability for the on-site system,
(2) system design, (3) installation, (4) operation, (5)
maintenance, and (6) rejuvenation of failing systems.
Facilities should not be permitted in areas where soil type,
subsurface conditions, groundwater conditions, or proximity
to water indicate a likelihood of malfunction or danger of
water pollution. In Maine, the state plumbing code requires
that a certified soil scientist conduct site investigations
to determine septic tank suitability.[11] Providing for proper
system operation and maintenance might include licensing of
plumbers and sewage scavengers, requirements for
documentation of septage disposal, and the provision of
septage disposal sites.[12]

Regulation of Solid Waste Disposal Sites

Standards for the establishment, location, operation,
maintenance, use, and discontinuance of sites and facilities
are often contained in state regulations. Watershed
management programs, however, should assure that such
regulations are designed to keep solid waste disposal sites
out of water supply watersheds and that they are enforced
locally. Regulations should also prohibit dumping along
rural highways in the watershed. This is often a problem in
exurban watersheds where people are living outside garbage
pick-up and trash pick-up areas. Conveniently located
dumping areas located outside the water supply watershed
would help alleviate the problem.

Regulation of Wastewater Discharges

Often the state must approve any municipal or industrial waste discharge in areas upstream from or affecting a public drinking water supply and the same statute or another statute prohibits anyone from damaging a public drinking water supply. These statutes might be enforced locally by local health departments. A watershed management program should assure that this aspect of protection is fully and forcefully implemented by providing for a monitoring and surveillance capacity. For Sebago Lake in Maine, the Portland Water District employed a full-time inspector who worked with town code enforcement officials and individual property owners. He also responded to complaints and walked every foot of the shoreline every year looking for signs of malfunctioning sewage disposal systems. State law gives the district power to inspect and approve all structures within 200 feet of the lake and to inspect drainage and sewage disposal systems within 1,000 feet.[13] Such a person could also seek out and initiate prosecution of persons or businesses responsible for dumping liquid or solid waste on land surfaces or directly into the storm sewers and receiving waters or for discharges in excess of permits.

Regulation of Impoundment Surface and Shoreline

Any water supply protection program should include the management of the reservoir surface and its shoreline. Shoreline regulations should stipulate a shoreland zone covering all land owned by the water purveyor and, if neccessary, beyond. Such zones, ranging up to 1,000 feet from the high water elevation, should be handled differently from the rest of the drainage basin. The regulation should limit tree cutting and shrubbery clearing, earth movement, construction, recreation and other activities, operation of vehicles, and waste material handling in the zone.[14]

Use of the reservoir waters should also be controlled, of course. Shoreland controls resolve some problems by controlling adjacent land use and access to the impoundment, but a water surface control strategy should be designed along with the shoreland control strategy. The practices to be controlled are suggested in Chapter IV.*

_____

*Also see reference 15.

ACQUISITION OF PROPERTY RIGHTS: APPLICATION OF COMPENSATORY POWERS

The second general category of governmental powers is the power to pay compensation for property being affected by watershed management. Often referred to as land acquisition or purchase of property rights, compensation enables restrictions on watershed uses and practices that go beyond legal limits of the police power. The cost to the public is correspondingly higher, of course.

The power to acquire land may be expressly granted to municipalities and other local authorities by specific acts of the state legislature. Alternatively, it may be implicit where it is necessary to the exercise of specifically conferred powers or to powers essential to the purpose for which the local government or agency is created. Thus some statutory authority is required, just as in the exercise of regulatory powers. Usually a local government may acquire land or rights involved with land by voluntary purchase, gift, dedication, or condemnation. The power to acquire property by condemnation, i.e., through eminent domain, is more limited than the power to acquire it by other means and must generally be explicitly provided for by state enabling legislation.

Of course, the acquisition of land must be for a public purpose. In some states this may be interpreted more stringently than in others. In those states the power to acquire land will be limited to land actually to be used or employed by the public; the so-called "use by public" test. Thus acquisition of land for an impoundment might be permissible, but purchase of land nearby in order to keep development away from the water supply might not be in those states. In other states, a more liberal interpretation of public purpose, called the "public benefit" test is in effect. There, an acquisition would be permissible as long as it tends to promote the welfare of the community.

At one time complete acquisition of entire water supply watersheds was a viable option for protecting drinking water quality. In general, massive land purchases are no longer feasible because of high land values and political opposition. Instead, communities must make strategic decisions about how to spend limited funds available to purchase watershed property rights.

Full-Fee or Fee Simple Acquisition

Fee simple acquisition of property is useful where the severity or selective application of a desired restriction on private land use rules out police power regulations. It has been proposed as a last resort where voluntary soil conservation practices will not be adopted by landowners and where the necessary limitation on land use without compensation would involve an unconstitutional taking. Condemnation of private property, i.e., the use of eminent domain, for use in promoting a desired urban development pattern is restricted by state constitutions which frequently specify that such powers can only be used by localities to provide necessary public services. Therefore, flexible acquisition methods such as installment buying, options, right of first refusal, and purchase and lease-back, must often be used in negotiating purchase prices and other terms with landowners who are willing to sell.[16] For example, several large tracts next to the South Fork Rivanna Reservoir serving Charlottesville, Virginia, were bought by working closely with the Nature Conservancy and securing funds from the state's open space program.[17]

Easements

On lands that should be protected from some forms of development but not all, and which require no public access, acquisition of negative easements that restrict certain private uses are an option in lieu of fee simple purchase. This measure involves paying private property owners a portion of the market value of their land in return for their agreement not to develop their property for certain uses that threaten water quality. It provides nonpossessory, less-than-fee simple interest in the subject land. Advantages of utilizing easements to limit development include: (1) land remains in private ownership and is usually maintained by the owner at no public expense while generating property tax revenue; (2) cost is generally less than fee-simple acquisition; (3) surrounding property values often increase, off-setting the tax loss on restricted property; (4) funds can be spread further to protect more land than with outright acquisiton; and (5) land use options are provided for the future.[18] Of course, easements do not provide as much control over the land as fee simple purchase.

Selling an easement, in lieu of selling the land outright, can also be of benefit to the landowner. He or she can remain on the land and make use of it in ways not inconsistent with terms of the easement. There may be tax advantages provided by preferential taxation schemes or simply by the fact that the limitations on use reduce the appraised value of the land. If the easement is a gift, it could qualify as a deduction from taxable income as a charitable contribution. Frequently, landowners may be persuaded to make bargain sales or dedicate easements in order to receive concessions from local government on development regulations or achieve tax benefits.[19] For example, trusts funded from private and public sources were used in Baltimore County and in Marin County, California, to pay farmers not to develop their properties.[20]

Transfer of Development Rights

TDR, as this approach is called, allows the acquisition of development rights in environmentally critical parts of the watershed, or conceivably all of the watershed, by the private sector rather than the public. The basic concept underlying TDR is similar to that underlying the use of easements--that ownership of land is a bundle of rights, each of which may be separated from the rest and transferred to someone else. The right to develop is one of those rights, as are mining rights or air rights. TDR, however, allows transfer of a detachable development right to another piece of land, preferably to a particularly suitable piece of land outside the watershed where development would not be detrimental to the quality of the raw water supply.

In the TDR approach, the watershed or particularly vulnerable sections of it are designated as restricted development or conservation zones, where use of the land is severely constrained. The landowner there is allowed to sell the unuseable development rights to a landowner in a receiving zone designated as suitable for development more dense than would otherwise be allowed by zoning and other land use controls. Development at the increased intensity in the receiving zone requires the developer of that land to have purchased the necessary development rights from a landowner in a conservation zone. Government may provide a market for such rights, maintaining a revolving bank of development rights, purchasing them from landowners in the watershed and recouping costs by making the rights available for purchase by landowners and developers in receiving zones outside of the watershed.[21]

Although being discussed here under the heading of compensation, TDR is a hybrid approach combining regulations by which restrictions are placed on the conservation zone land and compensation by which restricted landowners are compensated for loss of their development rights. The cost of compensation is shifted, however, from the public to the private sector. While the authors are unaware of any use yet of TDR to protect drinking water supplies, local governments are beginning to experiment with the concept and it is potentially suitable for watershed management. TDR has been proposed for a water supply watershed in Marin County, California.[22]

PUBLIC INVESTMENT IN CAPITAL IMPROVEMENTS

The power to build community facilities, including development serving facilities such as water and sewer lines, is closely related to the power to acquire property rights. Both are included under government's power to spend. For purposes of using public investments to protect the water supply watershed, it is the power not to spend that is relevant. That is, it is the withholding of water and sewer facilities from watersheds in order to discourage urban development there that is most applicable to watershed protection strategies.

The existence of public sewer and water, roads, schools, and other community facilities and services are key elements in private development decisions. The management of the location and timing of capital improvements therefore provides another method for protecting water supplies. It is an indirect land use control that schedules new roads, water lines, sewerage, and other public facilities in locations where development that is made possible thereby will not degrade water quality. These development-encouraging facilities may be proposed for locations out of the watershed altogether.

Development location and timing can be controlled using basic service districts to assure that denser residential development as well as industrial and commercial development will occur where public services are available, outside the watershed or at least not in particularly vulnerable parts of the watershed.

Regulations that require the existence of adequate facilities can be used in conjunction with the designation of service districts and the programming of those capital improvements to limit the amount and location of "urban" development.[23] In Baltimore County, basic service district maps are used to assure that urban development will occur where public services are adequate to handle it without adverse impact on water quality.[24] The ordinance might allow the developer to supply the capital improvements that would protect the water supply in lieu of withholding development permits in the absence of publicly constructed improvements. For example, again in Baltimore County, a shopping center developer was required to install a large detention basin and filters to treat the first two inches of runoff from parking lots.[25]

When a local government controls public services to manage growth, it generally coordinates its public improvement program with its comprehensive land use plan. Often, however, special districts or private utilities provide sewer, water, and other utilities. Therefore, intergovernmental coordination will be required because, while a municipality or county may use zoning to control some activities within a special district, express statutory authority is required to use the provision of water and sewer to control growth since the special district is also considered to be a local government.[26]

Capital improvements planning is long range in nature and requires careful consideration of current and future fiscal resources that will be needed to complete public utility expansions called for by such a plan. If used correctly, however, this form of planning may avoid some of the political heat that is generated by enforcing land use ordinances that restrict development in areas where water, sewer, and roads are present and water quality protection is the only basis for restricting growth. On the other hand, if development occurs anyway without adequate public services, particularly sewerage, the strategy of withholding services may cause severe water quality problems. A case in point: in Albemarle County, Virginia, plans are under way to extend an interceptor line after the fact to collect sewage from several industries and small residential communities that are polluting the water supply.[27]

A more direct and equally important use of capital improvements is in the construction of stormwater control and treatment facilities to control hydrologic processes and to treat raw water before distribution as potable water. Virtually all water systems must make such capital investments and they should be conceived of as integral elements of a watershed management program.

PREFERENTIAL TAXATION

Taxation is the fourth governmental power to be considered in a watershed protection strategy. The primary purpose of taxation is, of course, to raise revenue, not to

183

control land use. For this reason there are constraints on departures from the revenue raising function. Nevertheless, taxation, particularly the property tax, can have significant impact upon land use decisions. Thus, differential taxation can be used to provide incentives and disincentives for particular land uses and development practices. Taxation should not be the linchpin of a watershed management program. Its function is to complement regulations, the acquisition of property rights, and the controlled provision of community facilities.

Legal constraints upon the use of differential taxation focus mainly on the issues of substantial equity and uniformity of treatment. Classification of properties or persons must not be unreasonable or arbitrary. All persons similarly situated must be treated similarly throughout the watershed and even across the county and throughout the state. As with other local governmental powers, there must be statutory authority for the use of taxation in a watershed management program. That is, the state must allow the purpose for which the tax scheme is being employed. The state must also have given the power to impose the tax, explicitly or implicitly, to the particular government employing the differential taxation scheme. Generally this is the county or municipality rather than a special purpose government, such as a water district. Thus, the use of differential taxation, for example in a watershed management program, generally requires intergovernmental cooperation.

Most states have provisions allowing local taxing entities to assess agricultural and in some cases other open space land at its value for agricultural or other open space uses rather than its value for urban development. The major purposes of the state legislation are to help preserve agricultural land and help the farmer by reducing taxes in return for restriction on the use of the land. However, to the extent that at least the more benign agricultural uses in the water supply watershed have less adverse impact on water quality than urban development, preferential taxation can also be an appropriate element in a water supply protection program. Baltimore County, for example, utilizes preferential taxation in the water supply watershed to reduce the pressure on farmers to sell land in order to pay current taxes.[28]

When a landowner agrees to the use-value assessment, an agreement that provides for the payment of deferred taxes if the property is sold at its market value is normally signed. State laws vary, but most require payment of five to ten years' deferred taxes. In addition, some states require that preferentially taxed land be subject to some enforceable use limitations, such as deed restrictions.[29]

It has been found that preferential taxation will not work unless it is backed by strong public relations and enforcement of agreements, once signed. Second, it is voluntary and relies on the landowner to elect participation. Third, it does not offer permanent protection since it allows for eventual transition to urban development. Preferential taxation simply delays such a transition by reducing but not eliminating the pressure to convert to urban development. Thus, use-value assessment should not be viewed as a cure-all.[30]

It should be noted that, in addition to preferential taxation, a large number of tax incentives exist at the federal and state level for landowners who protect environmentally sensitive lands. Tax reductions can result from donations, bargain sales, life-estate transactions, gifts of conservation easements, and corporate donations.[31] Localities could take steps to counsel watershed landowners on their tax options as part of a management program.

OTHER COMPONENTS OF A WATERSHED MANAGEMENT PROGRAM

In addition to regulations, acquisition, capital improvements, and taxation, there are other measures that a water supply system might consider.

Land Surface Sanitation Practices

For urbanized parts of the watershed, the municipality or water district should consider measures for keeping the land and storm drainage system clean so that stormwater runoff will contain less pollution. These include:

1. Street sweeping improvements—for reducing accumulations of urban street litter; vacuum sweeping and removal methods rather than sweeping-and-washing of pavements and gutters are strongly preferred in order to remove pollutants from the watershed. Street surface contaminants represent a major portion of urban land pollution.[32] In order to increase effectiveness, particularly for smaller particles, a street sweeping program should include better training of cleaning equipment operators, proper maintenance of the equipment, and a decrease in operation speeds together with multiple passes.[33]

2. Cleaning catch basins and storm sewers of accumulated
    debris. Again, vacuuming and removal from the
    watershed is recommended rather than washing it away.
    When frequently cleaned, catch basins are effective
    in removing solids, including some of the fine
    particles, from stormwater runoff.[34/35/36]

Centralized Management of the Operation and Maintenance of On-site Disposal Systems

Improper use and inadequate maintenance are primary
causes of septic tank failures. This is especially true of
cluster septic systems, where no one individual or owner of a
unit can be held responsible for the failure of an entire
system. Better performance of conventional and alternative
on-site facilities can be expected through centralized
operation and maintenance services provided through a
governmental unit. Operation and maintenance services should
probably include the following: periodic inspections,
septage pumping, maintenance and repair of equipment,
absorption field repair and replacement, grounds maintenance,
and system change or alteration.

Not every local governmental unit in every state will
have clear legal authority to provide on-site wastewater
disposal services. The use of such authority should be
accompanied by appropriate power to support the services from
fees and charges, to borrow money, and to own and operate
facilities. Ownership of facilities is especially
appropriate for the operation and maintenance of cluster
systems or package plants serving subdivisions, mobile home
parks, motels, rural schools, and the like.

Encouraging Voluntary Land Management Practices by Farmers

In the absence of regulations, most communities must
rely on voluntary approaches for controlling pollution from
farms. In many areas cost sharing arrangements are available
as an incentive for private land owners to adopt pollution
abatement measures. Funds for cost sharing programs come
almost entirely from the federal government, which has
usually paid up to 50 percent of the expense of installing
pollution control devices.* In some instances, local governments have provided funds to increase the amount of money

---

*Under some cost-sharing programs funded by the U.S.
Department of Agriculture, up to 75 percent of the expense of

available for cost sharing programs. In addition, some communities have tried to get landowners to sign watershed protection agreements which commit them to control nonpoint pollution coming off their land as a condition for receiving a permit for development on their property.[37]

Public Relations, Education, and Participation

Public involvement through well designed information, education, and participation programs is vital to a successful watershed management program. For example, the Portland (Maine) Water District has distributed maps and literature about the importance of Sebago Lake as a water supply for over twenty-five years. Easily readable charts comparing water quality in restricted areas and non-restricted parts of the lake have been distributed. Also, inspectors and other water district personnel are trained to take a friendly and helpful attitude in enforcing watershed regulations. Finally, its 208 plan recommended establishment of a watershed planning association, which would offer advice on protecting the lake.[38] The Norfolk Watershed Management Program takes such an approach as well in a plan to launch an extensive public education campaign, including television and radio spots, workshops, and newspaper features.[39]

In a study of alternative policies for controlling non-point agricultural sources, Seitz et al. found that farm community acceptance of agricultural control practices is better if traditional agricultural agencies are involved in explanation and implementation of the policies.[40] Implementation of Best Management Practices in rural areas, there-fore, should include participation of traditional agencies such as the Soil Conservation Service and County Extension Services.

Legal Action

When various locational and site level controls are ineffective, communities can resort to legal actions to protect their water supplies. Threats to water quality can be addressed either through administrative action or court suits. Administrative actions involve protests or other actions taken through government channels when existing regulations are not properly enforced. Court suits are used

-----

soil erosion and sedimentation control devices may be paid to landowners.[41]

when either a government agency or a land user continues to
ignore regulations or when a land use action endangers public
water supplies. Although legal actions are not always
successful, they do increase public scrutiny and help focus
attention on the need to protect water supplies.

Portland Water District filed suit to stop a developer
from constructing a resort development on an island in Sebago
Lake when it appeared it might endanger the water supply.
Although development was not prevented altogether, the suit
did result in a lesser scale development and increased public
surveillance of its impacts.[42/43]

SUMMARY

Several basic powers and a variety of tools exist for
water supply watershed management. The formulation of a
management program calls for the creative application of
these basic powers and specific measures. That is, intervention measures for a particular case should not just be
selected off the shelf. They should be custom designed to
fit the political, developmental, and hydrologic characteristics of the watershed and the particular goals of the local
governments involved. Specific intervention measures are
usually custom designed applications of the basic powers as
they are defined in the particular state and locality. They
should not just be copied from another watershed program.
Thus, not only should the design team contain a planner and a
person knowledgeable in hydrology and water resource technology, but it should also include an imaginative attorney
and a creative and supportive administrator. Further, the
planner and technical expert should understand basic governmental powers and have some knowledge of the range of
measures that can serve as the source of ideas.[*]

The effectiveness of the measures briefly described in
this chapter depends on whether they are appropriate for the
particular watershed being managed, whether they are suitably
adjusted to fit the specific situation, and whether they are

---

[*]The most appropriate references to be used in exploring
the nature of the basic powers and the legal considerations
can be recommended by a city or county attorney in the
locality. As initial general references, references number
44 and 45 are recommended. Also valuable will be a
compilation of relevant statutes and court decisions in the
state, which a practicing planner or attorney in the locality
should have.

properly implemented and monitored. Chapters VI and VII
address the process of designing, implementing, and
evaluating a management program for a particular watershed.

REFERENCES

1. Miller, Todd L. and Raymond J. Burby with Edward J.
    Kaiser and David H. Moreau. Protecting Drinking Water
    Supplies Through Watershed Management: A Casebook for
    Devising Local Programs, Center for Urban and Regional
    Studies, The University of North Carolina at Chapel
    Hill, Chapel Hill, N.C., August 1981, p. 22.

2. Miller, Todd L. and Raymond J. Burby with Edward J.
    Kaiser and David H. Moreau. Protecting Drinking Water
    Supplies Through Watershed Management: A Casebook for
    Devising Local Programs, Center for Urban and Regional
    Studies, The University of North Carolina at Chapel
    Hill, Chapel Hill, N.C., August 1981, p. 114.

3. Miller, Todd L. and Raymond J. Burby with Edward J.
    Kaiser and David H. Moreau. Protecting Drinking Water
    Supplies Through Watershed Management: A Casebook for
    Devising Local Programs, Center for Urban and Regional
    Studies, The University of North Carolina at Chapel
    Hill, Chapel Hill, N.C., August 1981, p. 25.

4. Miller, Todd L. and Raymond J. Burby with Edward J.
    Kaiser and David H. Moreau. Protecting Drinking Water
    Supplies Through Watershed Management: A Casebook for
    Devising Local Programs, Center for Urban and Regional
    Studies, The University of North Carolina at Chapel
    Hill, Chapel Hill, N.C., August 1981, pp. 25, 60-61.

5. Thurow, Charles, William Toner, and Duncan Early.
    Performance Controls for Sensitive Lands: A Practical
    Guide for Local Administrators, U.S. Environmental
    Protection Agency, Washington, D.C., 1975.

6. Brown, William M., Walter G. Hines, David A. Ricket, and
    Gary L. Beach. A Synoptic Approach for Analyzing Erosion
    as a Guide to Land-Use Planning, Circular No. 715-L,
    U.S. Geological Survey, Reston, Va., 1979.

7. Blackman, Douglas and David W. Blaha. "Guidebook for
    Protecting the Quality of Surface Drinking Water
    Supplies in the Southern Piedmont," Master's project
    submitted in partial fulfillment of the requirements for

the Master of Environmental Management in the School of Forestry and Environmental Studies, Duke University, Durham, N.C., 1981, pp. 34-35.

8. Miller, Todd L. and Raymond J. Burby with Edward J. Kaiser and David H. Moreau. Protecting Drinking Water Supplies Through Watershed Management: A Casebook for Devising Local Programs, Center for Urban and Regional Studies, The University of North Carolina at Chapel Hill, Chapel Hill, N.C., August 1981, pp. 269-74, 289-95.

9. Miller, Todd L. and Raymond J. Burby with Edward J. Kaiser and David H. Moreau. Protecting Drinking Water Supplies Through Watershed Management: A Casebook for Devising Local Programs, Center for Urban and Regional Studies, The University of North Carolina at Chapel Hill, Chapel Hill, N.C., August 1981, pp. 63-65, 85-96.

10. Triangle J Council of Governments. Summary of Alternative On-site Wastewater Treatment and Disposal Methods, Triangle J Council of Governments, Research Triangle Park, N.C., November 1978.

11. Miller, Todd L. and Raymond J. Burby with Edward J. Kaiser and David H. Moreau. Protecting Drinking Water Supplies Through Watershed Management: A Casebook for Devising Local Programs, Center for Urban and Regional Studies, The University of North Carolina at Chapel Hill, Chapel Hill, N.C., August 1981, p. 213, Table VI-3.

12. North Carolina Department of Natural Resources and Community Development. Draft Water Quality Management Plan for On-site Waste Disposal, Raleigh, N.C., 1979.

13. Miller, Todd L. and Raymond J. Burby with Edward J. Kaiser and David H. Moreau. Protecting Drinking Water Supplies Through Watershed Management: A Casebook for Devising Local Programs, Center for Urban and Regional Studies, The University of North Carolina at Chapel Hill, Chapel Hill, N.C., August 1981, p. 210-217.

14. Lesniak, John. "Land Use Controls for Optimum Development of Federal Reservoirs in North Carolina," Report to the Department of Natural Resources and Community Development, Center for Urban and Regional Studies, The University of North Carolina at Chapel Hill, Chapel Hill, N.C., August 1977, pp. 22-34.

15. Miller, Todd L. and Raymond J. Burby with Edward J.
     Kaiser and David H. Moreau. Protecting Drinking Water
     Supplies Through Watershed Management: A Casebook for
     Devising Local Programs, Center for Urban and Regional
     Studies, The University of North Carolina at Chapel
     Hill, Chapel Hill, N.C., August 1981.

16. Schwartz, Seymour I., Robert A. Johnston, James R.
     Blackmarr, and David E. Hansen. Controlling Land Use for
     Water Management and Urban Growth Management: A Policy
     Analysis, California Water Resources Center, University
     of California, Davis, Calif., 1979.

17. Miller, Todd L. and Raymond J. Burby with Edward J.
     Kaiser and David H. Moreau. Protecting Drinking Water
     Supplies Through Watershed Management: A Casebook for
     Devising Local Programs, Center for Urban and Regional
     Studies, The University of North Carolina at Chapel
     Hill, Chapel Hill, N.C., August 1981.

18. Toubier, Joachim and Richard Westmacott. Water Resources
     Protection Measures in Land Development--A Handbook,
     Water Resources Center, University of Delaware,
     Wilmington, Del., 1974.

19. Smith, Clyn. "Easements to Preserve Open Space Land,"
     Ecology Law Quarterly, 737 (1971).

20. Miller, Todd L. and Raymond J. Burby with Edward J.
     Kaiser and David H. Moreau. Protecting Drinking Water
     Supplies Through Watershed Management: A Casebook for
     Devising Local Programs, Center for Urban and Regional
     Studies, The University of North Carolina at Chapel
     Hill, Chapel Hill, N.C., August 1981.

21. Schwartz, Seymour I., Robert A. Johnston, James R.
     Blackmarr, and David E. Hansen. Controlling Land Use for
     Water Management and Urban Growth Management: A Policy
     Analysis, California Water Resources Center, University
     of California, Davis, Calif., 1979.

22. Miller, Todd L. and Raymond J. Burby with Edward J.
     Kaiser and David H. Moreau. Protecting Drinking Water
     Supplies Through Watershed Management: A Casebook for
     Devising Local Programs, Center for Urban and Regional
     Studies, The University of North Carolina at Chapel
     Hill, Chapel Hill, N.C., August 1981.

23. Kuhner, Jochen, Russell deLulia, and Michael Shapiro. "Assessment of Existing Methodologies for Evaluation and Control of Watershed Land Use in Drinking Water Supply Systems," in Drinking Water Quality Enhancement Through Source Protection, Robert B. Pojasek, ed., Ann Arbor Science Publishers, Ann Arbor, Mich., 1977, pp. 345-374.

24. Miller, Todd L. and Raymond J. Burby with Edward J. Kaiser and David H. Moreau. Protecting Drinking Water Supplies Through Watershed Management: A Casebook for Devising Local Programs, Center for Urban and Regional Studies, The University of North Carolina at Chapel Hill, Chapel Hill, N.C., August 1981, pp. 26, 63.

25. Miller, Todd L. and Raymond J. Burby with Edward J. Kaiser and David H. Moreau. Protecting Drinking Water Supplies Through Watershed Management: A Casebook for Devising Local Programs, Center for Urban and Regional Studies, The University of North Carolina at Chapel Hill, Chapel Hill, N.C., August 1981, p. 30.

26. Schwartz, Seymour I., Robert A. Johnston, James R. Blackmarr, and David E. Hansen. Controlling Land Use for Water Management and Urban Growth Management: A Policy Analysis, California Water Resources Center, University of California, Davis, Calif., 1979.

27. Miller, Todd L. and Raymond J. Burby with Edward J. Kaiser and David H. Moreau. Protecting Drinking Water Supplies Through Watershed Management: A Casebook for Devising Local Programs, Center for Urban and Regional Studies, The University of North Carolina at Chapel Hill, Chapel Hill, N.C., August 1981, pp. 30, 309-10.

28. Miller, Todd L. and Raymond J. Burby with Edward J. Kaiser and David H. Moreau. Protecting Drinking Water Supplies Through Watershed Management: A Casebook for Devising Local Programs, Center for Urban and Regional Studies, The University of North Carolina at Chapel Hill, Chapel Hill, N.C., August 1981, p. 28.

29. Environmental Protection Agency, Council on Environmental Quality. Untaxing Open Space: An Evaluation of the Effectiveness of Differential Assessment of Farms and Open Space, U.S. Government Printing Office, Washington, D.C., 1976.

192

30. Gloudemans, Robert J. "Use Value Farmland Assessments: Theory, Practice and Impact," Studies in Property Taxation, International Association of Assessing Officers, Chicago, Ill., 1974.

31. Heritage Conservation and Recreation Service. Land Conservation and Preservation Techniques, U.S. Department of the Interior, Washington, D.C., 1979.

32. Sartor, J. and G. Boyd. Water Pollution Aspects of Street Surface Contaminants, EPA R2-72-081, U.S. Environmental Protection Agency, Washington, D.C., 1972.

33. U.S. Environmental Protection Agency. Water Quality Management Planning for Urban Runoff, EPA 440/9-75-004, Office of Water Planning and Standards, U.S. Environmental Protection Agency, Washington, D.C., December 1974.

34. Blackman, Douglas and David W. Blaha. "Guidebook for Protecting the Quality of Surface Drinking Water Supplies in the Southern Piedmont," Master's project submitted in partial fulfillment of the requirements for the Master of Environmental Management in the School of Forestry and Environmental Studies, Duke University, Durham, N.C., 1981, pp. 25-25.

35. American Public Works Association. Water Pollution Aspects of Urban Runoff, WP-20-15, Federal Water Pollution Control Administration, Washington, D.C., 1969.

36. Athayde, D. Preventive Approaches to Stormwater Runoff, EPA 440/9-77-001, U.S. Environmental Protection Agency, Washington, D.C., 1977.

37. Miller, Todd L. and Raymond J. Burby with Edward J. Kaiser and David H. Moreau. Protecting Drinking Water Supplies Through Watershed Management: A Casebook for Devising Local Programs, Center for Urban and Regional Studies, The University of North Carolina at Chapel Hill, Chapel Hill, N.C., August 1981, p. 119.

38. Miller, Todd L. and Raymond J. Burby with Edward J. Kaiser and David H. Moreau. Protecting Drinking Water Supplies Through Watershed Management: A Casebook for Devising Local Programs, Center for Urban and Regional Studies, The University of North Carolina at Chapel Hill, Chapel Hill, N.C., August 1981, p. 218.

39. Miller, Todd L. and Raymond J. Burby with Edward J. Kaiser and David H. Moreau. _Protecting Drinking Water Supplies Through Watershed Management: A Casebook for Devising Local Programs_, Center for Urban and Regional Studies, The University of North Carolina at Chapel Hill, Chapel Hill, N.C., August 1981, pp. 343-345.

40. Seitz, W.D. and C. Osteen. "Economic Aspects of Policies to Control Erosion and Sedimentation in Illinois and Other Corn Belt States", in R. C. Loehr et al., _Best Management Practices for Agriculture and Silviculture_, Ann Arbor Science Publishers, Inc., Ann Arbor, Mich., 1978, pp. 373-382.

41. Haith, D.A., and R.C. Loehr. _Effectiveness of Soil and Water Conservation Practices for Pollution Control_, U.S. Environmental Protection Agency, Washington, D.C., 1979.

42. Miller, Todd L. and Raymond J. Burby with Edward J. Kaiser and David H. Moreau. _Protecting Drinking Water Supplies Through Watershed Management: A Casebook for Devising Local Programs_, Center for Urban and Regional Studies, The University of North Carolina at Chapel Hill, Chapel Hill, N.C., August 1981, p. 215-17.

43. Grady, Robert P. "Source Protection on a Multipurpose Lake: A Utilities' Perspective," in Robert B. Pojasek, 1977, pp. 575-77.

44. Hagman, Donald L. _Urban Planning and Land Development Control Law_, West Publishing Co., St. Paul, Minn., 1971, 1975.

45. Wright, Robert and Susan Webber. _Land Use in a Nutshell_. West Publishing Co., St. Paul, Minn., 1978.

CHAPTER VI

PROGRAM DESIGN AND IMPLEMENTATION

The design and implementation of watershed management programs must go hand in hand. There are three reasons. First, programs designed without an appreciation of the technical and financial capacities of local governments which must administer them are likely to encounter many roadblocks which will frustrate their implementation. Second, programs are more likely to be implemented successfully if they incorporate specific measures designed to build and sustain support from the public and elected officials. Finally, programs are likely to more effective if design and implementation are viewed as a revolving process in which the program is adjusted as new scientific data become available and political conditions change.

In this chapter, we describe a three-step process for formulating and implementing a water supply watershed management program. The first step is called "tooling up." It involves gathering together the staff support and financial resources needed to conduct the preliminary studies and analyses described in the preceding chapters and to move forward with a management program. The second step is called "choosing a course of action." This step involves three sets of analyses. One set produces data about the cost-effectiveness of the alternative management measures emerging from the preliminary problem analysis and direction setting activities described in Chapters III and IV. It indicates which program measures, among those discussed in Chapter V, will achieve the goals and objectives which have been established at least cost. The second set of analyses assesses the institutional basis for undertaking watershed management. In particular, it indicates whether various agencies have the legal authority to pursue the watershed management responsibilities which they may be assigned and how specific management measures can be formulated so as to withstand legal challenge. The third set of analyses produces data about the political feasibility of the alternative measures being considered. It indicates which are likely to garner more or less political support and suggests how measures might be modified to improve their chances of adoption. Based on these analyses, watershed planners and

program officials must choose a course of action which will:
(1) technically accomplish the goals of the program in an
efficient manner; (2) garner sufficient political support to
be enacted; and (3) survive possible court challenges. Once
a course of action has been charted, the third and final step
in the process we propose is to enact and administer the
various elements of the watershed management program.

TOOLING UP FOR A WATERSHED MANAGEMENT PROGRAM

The first step in undertaking a watershed management
program is to acquire the staff and financial resources which
will be needed to identify and analyze watershed problems,
establish objectives and program targets, formulate watershed
management measures, and guide them toward adoption. Once a
program has been initiated, staff and financial resources
must be adequate to administer the regulations and other
measures which have been put in place and to monitor their
performance. Although water systems will undoubtedly be the
first to upgrade their staffs and commit financial resources
to watershed protection, other agencies expected to play a
major role in whatever program is adopted will also have to
begin committing resources to watershed management at an
early point in the planning and management process.

Program Staffing

The case studies of eight watershed management programs
revealed that most water systems with successful programs had
added or reassigned personnel so that the time of at least
one individual was devoted almost exclusively to watershed
protection.[1] Although the person with lead responsibility
for the program was usually an employee of the water system
using the watershed as a source of supply, other approaches
were used as well. In the case of the Rivanna Water and
Sewer Authority in Virginia, for example, lead responsibility
rested with a "watershed management official" who was hired
by and reported directly to the Albemarle County manager. In
the case of the Delaware and Raritan Canal in New Jersey,
lead responsibility rested with the staff of the D & R Canal
Commission, a special agency created by the State of New
Jersey to look after the canal and its watershed. In Newark,
the city established the Newark Watershed Conservation and
Development Corporation to manage the city's 35,000-acre
Pequannock Watershed. As these cases illustrate, a number of
different staffing arrangements are possible, depending upon
local circumstances. The key lesson, however, is not that
many different organizational arrangements can be used for

watershed management, but that the need for watershed
protection was recognized and that staff with lead
responsibility for watershed management was allocated.

Once watershed management programs are underway, they
can be administered without an extraordinary commitment of
staff resources. In our national survey of 496 water system
managers, less than 10 percent reported devoting more than
one person full time (or the equivalent) to watershed
management. The average water system, as reported in Chapter
II, was devoting about four man-weeks per year to this task.
Because many water systems did not rate their programs as
very effective, this manpower loading should not be viewed
as a national norm to be emulated. It does indicate,
however, that watershed protection can be undertaken with
relatively minor additions to the number of water system (or
other lead agency) employees.

Staff Training

Water systems could pursue watershed management with
relatively modest staff commitments because they used
consultants to accomplish many of the more technical aspects
of problem identification and program formulation. However,
the use of consultants does not relieve water systems and
other agencies cooperating in a watershed management program
from seeing to in-house staff training needs. As shown in
Chapter II (pages 52-53), lack of appropriate professional
personnel, lack of needed technical information, and lack of
technical assistance from other agencies are all major
barriers to program effectiveness. These barriers can be
overcome with adequate staffing, as discussed above, and with
adequate staff training. At a very minimum, management
personnel should have a basic understanding of land use and
water quality relationships. The extent and orientation of
more comprehensive training depends upon the current level of
staff expertise, funds available for staff education, the
ability to use outside consultants for technical studies, and
the manner in which watershed management regulations will be
enforced.

Staff training for watershed management should be geared
to the same time table used to pursue water supply protection
objectives. For example, it is a waste of resources to train
agency personnel about site-level stormwater abatement
techniques when the management program is still in its
infancy and may take a totally different approach to
watershed management. In short, at the outset of management
efforts staff training should be general in nature and

197

oriented towards developing a basic awareness of management
alternatives. Then as the direction that the program will
take unfolds, staff recruitment and training should attempt
to develop the skills necessary for current design and
implementation activities.

Financing Watershed Management

The costs associated with a watershed management program
are borne by both the managers and the managed. It is
obvious that adequate funding is necessary to carry out
planning and management initiatives. However, the agency
which should pay for the program is not always as obvious.
Expenditures are normally paid by the taxpayers of the
jurisdiction administering a program or the water system's
rate payers. However, in watershed planning it will
frequently be important to consider cooperative funding
arrangements. This is particularly true when water is
exported from a watershed to be consumed elsewhere, as is
usually the case. In this situation, taxpayers of the local
government with the ability to manage the watershed (often a
rural county) do not directly benefit from expenditures on a
watershed management program, so that sharing of program
costs is appropriate as a matter of equity, as well as
political expediency.[*]

It is difficult to predict in the abstract what a
watershed management program should cost. At the outset of a
management program money will be needed to conduct special
studies and prepare land use plans. In some communities
hundreds of thousands of dollars have been spent in the
initial phases of watershed management efforts; however,
these costs can be lowered significantly when communities
take less sophisticated approaches to protecting water

---

[*]For example, in the city of Charlottesville and
Albemarle County, Virginia, an agreement was worked out by
which both jurisdictions share in the costs of managing their
water supply watersheds. First, an incorporated authority
was formed to handle water and sewer needs for both the city
and county thereby making water quality protection more than
just Charlottesville's concern. Then, extra revenues paid by
water customers were used to finance planning studies that
determined how best to protect water supplies. Finally, the
city and county agreed to jointly support a watershed
management official responsible for assuring adequate
enforcement of water quality protection regulations and
devising new ways to protect drinking water supplies.

supplies.* After the initial studies have been completed and a course of action is laid out, costs should decline substantially. In fact, in many communities watershed management initiatives can be implemented as part of existing land use management and building inspections programs, thereby reducing administration expenditures.

The amount of money a locality spends to gear up to manage its watersheds depends upon whether or not it is involved in water quality monitoring, compliance checks of detention basins and other site-level nonpoint pollution control structures, as well as computer assisted data storage and water quality modeling. Communities employing conventional zoning techniques and other less sophisticated planning approaches for managing watershed development will spend much less on their programs.

In addition to deciding how to fund government agencies involved in watershed management, it is necessary to consider how much money watershed inhabitants and landowners will be willing to devote to protecting water supplies. There is frequently tremendous pressure to gradually substitute side-payments for police power decisions.[2] This is particularly true for watershed management programs which can impose costs on people who do not benefit from water supply protection efforts, as suggested above.

Watershed landowners may ask that the public purchase easements rather than force landowners to provide natural buffers along stream corridors at their own expense. It is almost mandatory that cost-sharing money be appropriated to control agricultural pollution, since some states do not allow communities to enact pollution control ordinances for farmland. Unless adequate provisions are made to defer some of the costs that watershed landowners must pay, they are likely to organize to overturn costly management initiatives.

Personnel, both their number and training, and the potential financing which can be arranged for watershed management determine the scope and magnitude of the problem identification investigations undertaken at the start of the management process and the feasibility of various management measures which might be adopted to bring about desired watershed land uses and practices. The resources water systems and local governments are willing to devote to watershed management must be determined early in the planning process,

---

*See case studies in reference 1.

since they will act as a fundamental constraint on the choice of scientific procedures and the selection of management measures to be proposed for adoption. On the other hand, as understanding of the problem progresses, governments' willingness to contribute to watershed protection is likely to change. This reemphasizes the concept introduced at the beginning of this chapter that watershed management has to be viewed as a constantly revolving process.

CHOOSING A COURSE OF ACTION

At this point in watershed planning and management, a number of tasks have been accomplished. Watershed problems, both present and potential, have been identified. Goals and objectives have been formulated and agreement about them has been obtained from all of the major agencies which will or might be involved in a management program. Target pollutants, target sources, target locations, and target hydrologic processes have been identified. A number of management measures which will bring about the desired state of affairs in the watershed have been screened and a list of potential measures with some promise has been prepared. Preliminary discussions with agencies which might take part in the watershed management program have been held and a basic understanding of possible financial and staff resources for the program has been developed. Now it is time to choose the set of measures which will be proposed for adoption by the cooperating agencies.

The final selection of measures will depend on the outcomes of three interrelated analyses: cost effectiveness, institutional, and political feasibility. These should be conducted concurrently so that time is not wasted evaluating measures with little chance of adoption because they are not cost effective, are politically inappropriate, or are not legally defensible. Each analysis will contribute to the choice of management measures and so it is described in some detail here. At the start, however, we must emphasize that the final choice of measures will not be determined solely by the cost effectiveness, institutional, or political analysis results. Program formulation is more an art than a science and much will depend upon the experience and judgment of those assigned the job of proposing what should be done about watershed management.

Cost Effectiveness of Alternative Management Measures

Cost-effectiveness studies provide a basis for comparing and rank-ordering watershed management measures in terms of their contributions to program objectives and their relative costs. To fully appreciate the utility of cost-effectiveness evaluations, it is useful to understand how this technique was originally developed.

In the early 1970s, evaluators became uneasy about attempting to use cost-benefit analysis in devising social programs. Cost-benefit analysis is a technique developed to compare the costs and benefits of a proposed course of action in economic terms.[*] It requires that costs and benefits be translated into a common measure (usually monetary) and then compared, generally by computing either a benefit-to-cost ratio of net benefits, or some other value for summarizing the results of the analysis.

The specific application of quantitative analysis to the formulation of watershed management programs is extremely difficult. In fact, the U.S. Environmental Protection Agency's official position on using quantitative analysis to evaluate water quality protection plans is:

> No rigorous analytical method exists which will readily identify the best plan for an area...many factors should be considered in comparing alternatives. While some of the factors, in particular cost assessments, can be quantified, others can only be qualitatively assessed.... Plan assessment involves the comparison of all key factors deemed pertinent for reliable decision making.[3]

---

[*]The cost-benefit methodology was developed from economic theory and systems analysis. It is basically an accounting of the beneficial and adverse effects (benefits and costs) expected to flow from a project. Cost-benefit analysis was first applied in water resources planning. Its purpose is to formulate optimal investment projects. It is most often used to choose among a finite number of engineering alternatives for projects such as dams or power plants. The approach has been developed and refined for some thirty years. The development of the approach may be seen in the references 4, 5, 6, 7, and 8. Cost-benefit analysis has also been used to evaluate completed projects. For example, see references 9 and 10.

Although the basic concept of comparing costs and benefits is useful to consider when choosing among alternatives, it is simply not possible to obtain agreement on how to place a monetary value on all the costs and benefits associated with managing watersheds.

The key to using cost-effectiveness analysis in developing watershed management programs is to realize that it does not require that all costs and benefits be monetized. Instead of attempting to establish the worth or merit of any single management system, cost-effectivensss analysis is used to compare the relative efficiency of different management methods which are all aimed at achieving similar results. Thus using cost-effectiveness analysis programs with similar objectives are evaluated and the costs of alternative programs for achieving the same objectives are compared.

For these reasons, cost-effectiveness analysis rather than cost-benefit analysis is a more appropriate technique for making decisions about how to spend resources for watershed management. Cost-effectiveness requires quantifying program costs and impacts, but the impacts do not have to be monetized, only the costs. Impacts may be expressed in terms of actual outcomes. For example, the effect that different soil conservation practices will have on drinking water quality can be measured in terms of projected reductions in soil loss, rather than converting soil erosion rates to dollar figures. Similarly, other indicators of the effectiveness of management tools can be compared. The aim of cost-effectiveness analysis is to identify the least cost management program which will achieve minimum program objectives.

Program costs are all public and private resources that go into watershed management. The first step in calculating costs is to examine the program description in order to construct an exhaustive list of its ingredients. A partial list of some costs associated with two alternative watershed management programs is shown in Table VI-1.

In this example, the first program relies on one-acre residential zoning and twenty-five acre agricultural zoning to protect water quality. On the other hand, the second program has a structural orientation depending upon site-level pollution control, public water and sewerage, and land acquisition to protect drinking water supplies. New development is expected to go to other communities if it cannot locate in the watershed.

202

Table VI-1. Cost–Effectiveness Analysis of Two Alternative
Management Programs for a 10,000 Acre Watershed

Category	Low Density Program	High Density Program
Watershed Statistics (assuming full development under program)		
No. of Dwelling Units	3,000 d.u.	7,000 d.u.
Farm Acreage	5,000 ac.	2,000 ac.
Undeveloped Acreage	2,000 ac/	2,000 ac.
Residential Acreage	3,000 ac.	5,000 ac.
Stream Buffers	Not Needed	1,000 ac.
Annual Costs		
Personnel	$ 1,000	$ 35,000
Facilities	3,000	15,000
Meterials & Equipment	500	3,000
Lost Property Taxes	342,000	—
Public Water	—	73,500
Public Sewerage	—	219,000
Land Acquisition	—	10,000
TOTAL ANNUAL COSTS	$347,400	$355,000
Program Impacts		
Water Quality		
–Annual Watershed Soil Loss	3 tons/ac.	3 tons/ac.
–Annual Lake Sedimentation	1 inch	1 inch
–Annual Loss of Reservoir Storage	7MG/year	7 MG/year
–Risk of Accidental Chemical Spills	Low	Moderate
–Risk of Increased Pesticides, Nutrients	Moderate	Moderate
–Risk of Heavy Metals, Other Chemicals	Low	Moderate
Annual Housing Vacancy Rates	3 percent	5 percent
Average Rental Rates	$235 mo.	$225 mo.
Farmland Preserved	5,000 ac.	2,000 ac.
Savings on New Reservoir	$3,000,000	$1,000,000
Increased Construction Costs	$800/d.u.	$3,000/d.u.
Reduced Land Values	$10,000,000	—

Most fiscal impacts of watershed management can be assessed either by looking at a program's budget or by following standard procedures for estimating administrative expenditures. Reduced property tax revenues are calculated based on the extent to which the value of land and personal property is reduced. In this case, the value of land and personal property will actually increase under the second program since population densities and demand for land will increase under the high density program. Estimates of lost property taxes and reduced land values that are shown for the low density program do not represent actual monetary losses, but rather are based on the amount taxes and land values differ between the two programs.

Program impacts that are listed in Table VI-1 illustrate how they are considered in cost-effectiveness analysis. Water quality is expected to be about the same under both management approaches; however, there is more uncertainty over water quality under the high density program. Housing will be in shorter supply under the low density program, but not to such an extent that it will place severe hardship on the community. More farmland is preserved under the first program. The high density program will eventually require capital investments to provide adequate drinking water supplies to the larger population it allows; however, both programs increase the life of the existing reservoir and therefore save money. Increased construction costs are much higher under the second program because of its site-level engineering requirements. Finally, land values will be lower under the low density program because of larger lots and less demand for land; however, no substantial decrease in current land values is projected.

At this point in evaluating alternative management measures, a judgment must be made as to whether the costs associated with a preferred management measure are worth the benefits and other impacts it derives. There is no set procedure for making this determination, but local officials will have an easier time assessing the worth of watershed management alternatives if costs and impacts are accurately accounted.

Institutional Analysis

At the same time that the costs and beneficial impacts of potential management measures are ascertained, it also has to be determined whether existing governmental institutions are adequate to adopt and administer the measures and if so, whether they are politically feasible. In assessing institutional arrangements, which is described in this section, the powers and capabilities of five types of agencies must be considered in terms of the specific management measures being evaluated for inclusion in the watershed management program. These agencies include: (1) the water system; (2) soil conservation districts; (3) local governments; (4) regional agencies; and (5) state and federal agencies. Each type of agency has distinct powers and limitations which make it more or less suited to assume responsibility for various watershed management tasks.

Water Systems

One of the most important powers that water systems possess is that they are a single purpose agency charged with the responsibility of supplying safe drinking water to the public. Thus, they are the natural catalyst for motivating local concern and government action for protecting public water supplies. In pursuing watershed management objectives, water systems have several important powers that they may use.

First, they have the power of persuasion. By conducting sanitary surveys, water quality monitoring, and other technical studies, water systems can document threats to water supplies in order to convince public officials and water system customers of the need for watershed management. Moreover, since many private and public water systems pass their costs directly to water users through their rate base, they have the ability to raise funds to undertake watershed management activities.*

Second, water systems' ability to file legal suits against polluters of water supplies is another important power. Suits can be brought on several grounds. They are almost always successful when used to force violators of existing federal, state or local water quality control regulations to stop polluting water supplies. In addition to being directed against polluters, court actions may also be brought against government agencies that are delinquent in

---

*The Portland Water District in Portland, Maine provides a good example of effective use of information to motivate local and state governments to protect sources of drinking water. An aggressive land acquisition and monitoring program combined with educational activities has enabled the District to obtain regulatory support to prevent hazardous land uses that might degrade drinking water quality.[11]

enforcing pollution control standards.

Legal actions are also effective when directed against land uses that pose potential dangers to water supplies but which do not violate a specific land use regulation or water quality standard. Frequently a proposed land use can be challenged on the basis that it will endanger public health. In many instances, even if the water system does not get a clear judgment in its favor, the increased attention paid to water supply protection will bring changes in the proposed land use that enhances water quality protection measures.*

Soil Conservation Districts

A second public agency that may be included in watershed management efforts is the conservation district. They are legal subdivisions of state government, responsible under state law for conservation work within their boundaries just as townships or counties are responsible for roads and other services. Their boundaries are typically drawn along county lines, with a few being organized by watersheds. Although state laws vary, the purpose of districts is similar everywhere: to focus attention on land, water, and related resource problems; to develop programs to solve problems; and to enlist and coordinate help from all public and private sources that can contribute to accomplishing the district's goals.[12]

Conservation districts are uniquely equipped to help plan, manage, and implement water supply watershed management programs. Although few districts have exercised power to enforce conservation regulations, under their assigned responsibilities they have perfected working relationships with a host of federal, state and local agencies, institutions and groups. Thus, their experience and success as a coordinator may be a valuable asset to tap when attempting to get people and governments to work together to protect water supplies.

---

*The Portland Water District sued a developer who proposed to intensely develop an island in its water supply lake on the grounds that the project would degrade water quality. The District lost the case, but in the process won concessions from the developer who agreed to lower his proposed building densities and take other steps to protect water quality.[13]

206

Except in rare instances where state legislatures or Congress directly intervene to protect public drinking water supplies, the real authority and responsibility for management of water supply watersheds rests with local governments. Local governments may act independently or in cooperation with other governments to manage water supplies. However, whatever action they take, it is likely to be based on the police power they possess as granted to them by both the federal and their state's constitution.

Regulations of land use activities in water supply watersheds is possible because of local government police powers. Police power is defined as the power of local governments to pass laws for the protection of the health, welfare, safety, and property of its citizens.[14] There are several limitations which serve as parameters for evaluating the appropriate use of the police power. First, it must be used for a public purpose rather than the benefit of a private interest. Second, regulations must not violate the Equal Protection Clause of the Fourteenth Amendment of the U.S. Constitution which deems that a statute may not be arbitrary or discriminatory. Third, to justify interference with private uses of land on behalf of the public, it must appear that the method of control is necessary to protect public welfare.[15] Fourth, and finally, the management measures cannot completely restrict the use of private property without just compensation.[16]

Watershed management programs should be examined in terms of the four limitations on the use of police power. The first restriction, that regulations serve a public purpose, will rarely require extensive examination since the protection of an area's water supply is clearly for the benefit of the entire public and does not serve a private purpose. The other limitations, however, demand more consideration.

Equal protection challenges of watershed land use regulations may arise based on the argument that all people are not treated equally.[17] One type of challenge may appear in the context that a statute is arbitrary or discriminatory because it treats persons under the same or similar circumstances differently.[18] For example, a regulation that requires buffer strips along a stream must apply to all riparian landowners.[19] Watershed regulations should be evaluated to determine if they treat all landowners of a certain classification the same.

Another type of equal protection challenge that has been brought against watershed regulations is based on the contention that they are discriminatory against poor people. This was an issue in a court case involving a watershed plan for Newark, New Jersey's Pequannock Watershed. Zoning requiring minimum lot sizes of five acres was attacked on the grounds that the watershed was in a developing community and therefore provisions had to be made to assure an appropriate variety and choice of housing.* The judge in the case ruled that to limit housing to large lots simply because of environmental problems has not been approved by the New Jersey courts.** Although state laws may vary, an evaluation of how a watershed program affects low income people is appropriate to avoid the threat of an equal protection suit.

Challenges based on the argument that a statute does not bear a reasonable relationship to public health, safety, and welfare have been quite numerous in the area of land use regulation for water supply protection. As discussed earlier in Chapters I and III, uncertainties about land use and water quality relationships make it difficult to predict precisely the effect of land use regulations on water quality. The resolution of these cases hinges on the amount of evidence that is presented to demonstrate the need for the particular land use regulation or the probability that the restricted activity will contaminate water supplies. Programs should be evaluated to determine if a good faith effort has been made to conduct technical studies which are aimed at making watershed management activities scientifically creditable.

------

*Mt. Laurel, 174, 179, laid down the test for a developing community in New Jersey: (1) has a sizeable land area; (2) lies outside the central cities and older built-up suburbs; (3) has substantially shed rural characteristics; (4) has undergone great population increase since World War II or is now in the process of doing so; (5) is not completely developed; and (6) is in the path of inevitable future residential, commercial, and industrial demand and growth. The Mt. Laurel decision found that developing communities must make provisions for low to moderate income people to reside within their boundaries.

**Ecological and environmental considerations were also advanced in the Mt. Laurel case to justify large lot zoning throughout the township. The court pointed out that while such factors and problems were always to be given consideration in zoning, the danger and impact must be substantial and very real, not simply a vague threat to support exclusionary housing measures or preclude growth.

The fact that an ordinance has a reasonable relationship to a public goal does not preclude it from violating the Fifth Amendment prohibition against taking of private property without just compensation. The Fifth Amendment requires that when a regulation is too restrictive, the regulation must either be struck down or through eminent domain proceedings the landowner must be compensated for the loss of property rights. The distinction between a valid use of the police power, which does not require compensation, and an illegal taking, which requires compensation, is continually debated by the courts.

There are a variety of tests that courts use to decide if an unconstitutional taking has occurred, and these should be used to evaluate watershed regulations. The "diminution in value test" is used to determine if land use controls substantially or totally destroy the value of property and if no viable use of the land remains. Courts also use the balancing test in evaluating taking challenges. Under this test, the interests of private property owners are balanced against the interests of the public. If the benefit gained by society due to a regulation is outweighed by the loss sustained by private landowners, a taking has likely occurred and compensation will be required.[20]

A recent decision by the U.S. Supreme Court indicates how difficult it is to evaluate the constitutionality of land use regulations. In Penn Central Transportation Company v. City of New York, the city refused to approve plans for construction of a fifty story office building over Grand Central Terminal, which had been designated as a landmark under the Landmark's Preservation Law. In upholding the city's action, the Court acknowledged that it had been unable over the years to determine "any 'set formula' for determining when 'justice and fairness' require the economic injuries caused by public action be compensated by government, rather than remain disportionately concentrated on a few persons," and therefore taking challenges must be evaluated on a case-by-case basis.[21]

This decision built on an earlier ruling by the Supreme Court. In Goldblatt v. Hempstead,[22] the Court reviewed the validity of a safety ordinance that required an owner of a gravel and sand mine to refill his excavation because of the dangers it presented to developing neighborhoods in the area. In upholding the ordinance, the Court listed several factors in determining its reasonableness: (1) the nature of the menace against which the ordinance protects and the extent of the public interest in that protection; (2) the availability and effectiveness of less drastic protective steps; and (3)

209

the loss which the landowners will suffer due to the government's interference. With regard to the taking challenge, the Court stated that the fact that the landowner had been deprived of the most beneficial use of his land does not make the regulation an unconstitutional taking.

An historical overview of the Supreme Court's major rulings on the validity of land use regulations indicates that while, in earlier cases, the Supreme Court rigorously applied the federal constitutional requirements that no private property be taken without just compensation, in later cases, the Court became increasingly reluctant to interfere with local zoning and land use decisions. However, state courts in applying their constitutional provisions may apply stricter standards in determining when a taking has occurred.[23] To the extent that a state court invalidates a local land use regulation on adequate and independent state grounds, the state court ruling is not reviewable by the federal courts.[24] Thus, the considerations outlined by the federal courts are minimal guidelines for evaluating the uses of the police power in land use regulations. Each state may include factors in its review under its state constitution. Consequently, in designing a water supply protection program, the law of the individual state must be consulted carefully.

When the locality acts under its general police power to protect its water supply, it should evaluate its actions to ensure that they are legally defensible. It is important that local ordinances contain a statement of purpose and intent which makes clear that the government's actions are for a public purpose and for the protection of the public health, safety, and welfare. Because courts closely examine the legislative intent when construing a statute, the reason for enacting the regulation and the rationale for choosing a particular type of regulation over other alternatives should be unambiguously set forth in the law.

It is also exceedingly important that a locality has based its regulations on the most indepth studies and data that it can generate. The documentation of the governmental action with a substantial scientific foundation can help to fortify the regulation against challenges that: (1) the regulation is not for a public purpose; (2) that it is not reasonably related to a legitimate police power purpose; and (3) that it is an invalid taking because the individual property owner's harm from the regulation greatly outweighs any benefits derived by the public.

Once watershed management regulations are imposed, evaluations of their enforcement are needed to ensure that a community is not delinquent in its duty to protect public welfare. Moreover, as in the Newark court case referred to earlier, existing watershed management activities may be challenged on the grounds that they are inadequate and do not protect drinking water quality.[25]

Often the water supply watershed of a locality does not lie within the complete jurisdiction of that locality. This situation creates problems for a community trying to regulate land uses to protect the water supply. Local governments are creatures of the state and do not have authority outside of their corporate boundaries in the absence of a grant of authority from the state.[26] Thus, localities whose water supply watershed is not under their jurisdiction must try to prevent contamination of their water sources by encouraging intergovernmental cooperation with neighboring local governments.

To facilitate intergovernmental cooperation, local governments may pursue several options for protecting their water supplies through use of their police powers. A few alternatives a locality may consider include: (1) special districts; (2) inter-local agreements; (3) litigation; and (4) lobbying for new state legislation.

Special districts can transcend local boundaries and therefore can be uniquely suited for managing watersheds. The basic requirement for a special district is that it be a "municipal corporation"--that is, an autonomous legal entity created under state law with specific legal powers and geographic territory.[27]

Many states have enabling legislation that allows local governments to establish special districts which have regulatory power to protect water supplies. In general, however, special districts are not well-suited to pursuing multiple approaches to watershed management. The broader their scope of activity the more likely they are to infringe upon the functions of local governments which also have jurisdiction over the watershed. Another weakness of special districts is the frequent inclusion in state enabling legislation of stringent requirements for the adoption of regulations (such as a two-thirds and three-quarter majority vote of approval from landowners or land occupiers in the district before it can enact an ordinance).[28] Thus, it is normally extremely difficult for special districts to enact strict land use controls.

For these reasons, special districts should be established for watershed management only when there is a limited function that they can effectively serve. For example, a district could be established to assure consistent administration of watershed zoning that has been enacted by local governments with jurisdiction over a particular drainage basin. This may be a preferable alternative to having each government administer its own ordinance since it would help to avoid uneven application of standards and spread the liability to all member jurisdictions in the case of a lawsuit.

Inter-local agreements are another option local governments can pursue to manage watersheds. Local governments are "legal persons" and therefore can enter into contracts and agreements with one another to manage watersheds. While contracts are formal legal instruments which specify in detail the obligations of each party, local governments may also enter into less formal agreements which lack the specific standards of a contract. A memorandum of understanding may express mutual policies and intentions without creating legal obligations.[29]

Several examples of the utility of such agreements and contracts appeared in the case studies conducted for this book. In Virginia, a contract between the City of Charlottesville and Albemarle County established the Rivanna Water and Sewer Authority which has been instrumental in making watershed management a regional concern. A less formal approach has been tried in Baltimore and Baltimore County, Maryland, where a memorandum of agreement concerning proper watershed management has been signed by government agencies involved in managing watershed land use activities.[30]

It is possible for a locality to secure enabling legislation from its state legislature authorizing direct land use regulation of water supply watersheds that are outside of its corporate boundaries. Direct grants of authority provide a locality with an exceptionally strong basis with which to deal with extraterritorial land uses affecting its water supplies. The obvious advantages of this arrangement are that the locality gets direct control over its watershed, and it is not forced to rely on intergovernmental cooperation or litigation to prevent contamination of its drinking water. However, except in cases where immediate action is necessary to protect public health, it is unlikely that state governments will be willing to usurp the land use authority of one government for the benefit of another jurisdiction.

In City of West Frankford v. Fullop, infra., an Illinois statute authorizing cities to protect public water supplies in areas up to ten miles outside their corporate limits was challenged.[31] The Illinois Supreme Court held that this law provided adequate extraterritorial authority for municipalities to enact ordinances prohibiting land use activities that would pollute their reservoirs. The Court found that because of the statute, cities were not limited to purchasing land under their eminent domain powers in order to control hazardous watershed development.

Another option available to local governments is to control land uses by pushing for better enforcement of existing water pollution laws. This route was used by Norfolk, Virginia to stop pollution of its reservoirs resulting from feedlot operations.[32] The city threatened to sue the State Water Pollution Control Board if it did not enforce pollution control laws to stop its water supplies from being degraded. This alternative for protecting drinking water quality is only satisfactory if a state or federal law exists that will remedy the pollution problem. Most point sources of pollution are regulated.

When every other local government power (see Chapter V) has been exhausted in efforts to protect water supplies, localities may resort to litigation. Court suits are almost always successful when they are brought under a state or federal pollution control law which is not being adequately implemented or enforced. In addition, suits may be brought under a common law nuisance challenge if a threat to public health and welfare can be proven. These cases are less likely to succeed in court, but delays and extra public attention for watershed development that they create frequently improve the level of protection for drinking water supplies.

Regional Agencies

Most regional agencies that can play important roles in watershed management take the form of councils of governments, district planning commissions, or 208 water quality agencies. These agencies have no authority to regulate watershed land uses except in rare instances where state governments have granted them such powers. Regional agencies are primarily in a position to motivate member governments to become involved in watershed management by conducting studies of water quality problems and by commenting on the appropriateness of federal and state expenditures in watersheds

213

through the A-95 review process.*

Once local governments have decided to undertake watershed management activities, regional agencies can continue to
provide technical assistance in areas where individual
governments lack expertise or funding to carry out necessary
studies. In fact, cooperative arrangements between
localities that allow regional agencies to provide services
that would otherwise be duplicated on the local level are
efficient and money saving steps that can be taken to help
manage watersheds. Coordination of local programs and
assistance in raising revenues for watershed management are
major roles that regional agencies can play in protecting
water supplies.

State and Federal Agencies

Existing state and federal laws and programs provide
local governments with a wide range of options for obtaining
support for their watershed management (see Chapter II).
These legislative mandates range from direct regulatory
involvement in watershed management to incentives that

---

*After the responsible planning agency has prepared a
comprehensive plan for some government action, but prior to
official action, the plan is usually sent to other interested
parties, including related public agencies, agencies of
adjacent governmental jurisdictions, private organizations
and citizen groups for their review and comment. This
process provides the means by which issues or disagreements
regarding the plan are surfaced. It is the responsibility of
the planning agency to reconcile these comments with the
plan, incorporate changes, to respond to them, and to revise
the plan if necessary.

One official review and comment process has been
established under OMB's Circular A-95. Circular A-95
establishes procedures for the review and coordination of
many federal activities which concern state and local
government. Particularly for applications for federal
assistance and to a lesser degree for program outputs, such
as the comprehensive plan itself, reviews are conducted by
state and areawide "planning and development clearinghouses"
under the overall administration of the Federal Regional
Councils. Clearinghouse agencies are usually state planning
offices or regional planning agencies. Through this review,
conflicts with plans of other jurisdictions and with other
programs can be identified and reconciled.

provide funds and technical assistance for localities and private individuals to protect water supplies. These laws provide the basis for watershed management activities carried out by local agencies.

State enabling legislation that allows local governments to plan and regulate land use is fundamental to watershed management programs. This legislation varies from state to state, but typically covers preparation of land use plans, subdivision and zoning ordinances, sedimentation control regulations, and other laws administered by local governments. State constitutions control the amount of leeway that localities have in administering state-granted powers, varying from states granting localities broad "home-rule" powers to states that only allow localities to exercise those powers which are explicitly spelled out by state statutes.

Some states have enacted controls over water supply watershed land uses. For example, most require buffers around reservoirs to prevent contamination from waterfront development. In North Carolina, septic tanks within water supply watersheds must be placed on lots that are no smaller than 40,000 square feet. The state also prohibits hazardous waste handling facilities from locating within water supply watersheds.

Federal involvement in watershed management is quite diverse. Under the Federal Water Pollution Control Act Amendments of 1972, Congress greatly expanded the federal emphasis on water quality management planning, particularly coordination among water pollution control activities at the different levels of government. These amendments integrated regulation of point sources of pollution into a three-tiered structure for planning operations, which includes: (1) municipal facilities planning (water and sewage treatment plants under section 201); (2) areawide water quality management planning by regional agencies (section 208); and (3) basin planning by state governments (section 303(e)).

Much of the planning and actual construction of facilities called for by the amendments is now complete. Data collected, analyzed, and presented in plans are available for use by localities in watershed planning. In some instances, such as in Baltimore County, 208 water quality planning has been specifically directed to solving water supply quality problems.[33]

The Safe Drinking Water Act of 1974 charges states with primary enforcement responsibility for ensuring the quality

of public drinking water supplies. Although the Act has no
explicit requirements for planning, it focuses attention on
state governments for enforcing national drinking water
standards. This responsibility implies that states and local
governments will have to be involved in protecting the
quality of water supplies if they are to succeed in meeting
drinking water quality standards.

Funds administered by the U.S. Department of Agriculture
to combat soil erosion and control sedimentation is another
major area of federal involvement in watershed planning.
Through two agencies, the Soil Conservation Service and the
Agricultural Stabilization and Conservation Service, technical and financial assistance is provided to local governments, farmers, and other landowners interested in protecting
water quality. Six specific federal programs that relate to
erosion and sedimentation control and which can be applied in
water supply watersheds are described in Table VI-2.

Summary

The institutional evaluation serves two major purposes
in screening measures for inclusion in the watershed protection program. First, it indicates whether agencies—local,
regional, state, or federal—have the legal authority to
adopt and administer the measures being considered. Second,
it indicates whether the measures being considered—
particularly regulations of various sorts—meet U.S. and
state constitutional standards. Beyond these "checklist"
functions, the institutional analysis should also suggest new
approaches for pursuing watershed management measures.
Innovation is possible in a number of areas, including state
enabling legislation, formation of new corporate entities,
such as special districts, and in forging new relationships
among existing jurisdictions and agencies through intergovernmental contracts and agreements. Thus, this phase of
choosing a course of action should not only shed light on the
most appropriate watershed management measures, but should
also help suggest ways of putting the measures to work most
effectively.

Political Feasibility Analysis

The political feasibility analysis is designed to
indicate whether potential measures are likely to be adopted
and to retain enough support throughout the life of the
program so that it can achieve its objectives. Obviously,
the science in political science has not yet reached a stage
where highly reliable predictions can be made about the fate

Table VI-2. Federal Erosion and Sedimentation
Control Programs

I. Conservation Operation Programs: SCS helps landusers
through organizations known as soil and water
conservation districts. SCS disseminates information
about conservation measures needed on each kind of soil
that enables the landuser to use these soils without
erosion damage. In addition, technical assistance
provided through this program assists county officials,
developers, contractors, and builders by providing
information on the potential use limitations of
different kinds of soil. On-site assistance is also
provided to the landusers in making a conservation plan
that includes treatment necessary to protect the soil
from erosion.

II. Small Watershed Program: Through this program
assistance is provided for planning and cost-sharing of
watershed projects under Public Law 566. This
assistance focuses on established soil and water
conservation measures for private and public land
through the construction of dams and other water
control structures and upstream tributaries to ensure
effective water management. Small watershed projects
are limited to areas no larger than 250,000 acres, with
project purposes that incude: flood control, water
supplies, irrigation water, and recreation. The SCS
may enter into individual or district agreements based
upon conservation plans and may provide technical
assistance for planning and financial assistance in the
form of cost-sharing. In addition, the SCS cooperates
with other federal, state, and local agencies to
conduct river basin studies and watershed or flood
prevention demonstration projects.

III. Resource Conservation and Development Program: The
Food and Agriculture Act of 1962 (P.L. 703) allows the
SCS to assist local sponsors of rural-urban projects by
coordinating the assistance of other federal and state
agencies in meeting program objectives of developing
the economic base and protecting the environment. This
is accomplished by the development of land and water
resource plans for agricultural, municipal, or
industrial use and for recreation and wildlife.
Projects must be sponsored by local representatives and
the SCS on conservation measures for watershed
protection and flood prevention. In the sponsored
RC & D planning agencies, there are provisions for

217

Table VI-2 (continued)

cost-sharing of RC & D projects to control soil
erosion, protect fish and wildlife, provide rural water
supplies, mitigate flooding, control agricultural-
related pollution.

IV. River Basin Survey Program: This program is authorized
through P.L. 566 and enables the SCS to join in
cooperative studies between the USDA, state
governments, and other federal agencies such as the
Water Resources Council, in basin planning. The
objective of the program is to cooperate with state and
federal agencies in coordinating upstream watershed
projects; to identify water and land resource problems;
and to suggest ways for coordinating resources. Such a
cooperative study is being conducted for the Yadkin-Pee
Dee river basin and to date has consisted of inventory
studies of the basin resources.

V. Resources Conservation Act: The section of this act
which deals with Land Inventory and Monitoring Program
directs the USDA to make a comprehensive assessment or
appraisal of the nation's basic natural resources and
to help protect and improve them. The SCS has primary
responsibility to appraise the nation's soil, water,
and related resources and to develop a 5-year program
to guide conservation efforts and evaluate the
effectiveness of ongoing conservation programs. As
part of an effort to develop a National Soil and Water
Conservation Program, the SCS will work in cooperation
with citizen groups, conservation districts, and other
federal, state, and local agencies.

VI. Soil Conservation and Domestic Allotment Act: This
program is authorized under P.L. 74-461 and provides
authorization for the SCS to make cost-share payments
to farmers for implementation of specifically approved
soil and water conservation measures.

Source: Kathy Blaha, Jeff Cohen, Rhonda Evans, Dave
Merdinger, Todd Miller, Roger Schecter, Lorrie
Schmitt, and Terri Silverberg, "Soil Stability and
Erosion in the Pee-Dee Yadkin Basin and Other
Tributaries to Winyah Bay," Winyah Bay
Reconnaissance Study, The Conservation Foundation,
Washington, D.C., 1980.

of policy proposals, particularly over time. Nevertheless, we now know enough about program adoption and implementation to indicate the key factors which affect program outcomes-- both positively and negatively. This information can be used to arrive at educated guesses about the political feasibility of alternative management measures.

Here we briefly review four reasons for program success or failure. We suggest that program analysts consider potential management measures in relation to these factors, scoring measures subjectively on each factor to indicate the direction and strength of its possible effect on program adoption and implementation. The indicators may then be summed to produce a general idea of the overall feasibility of management measures and to compare the feasibility of alternative measures for achieving the same objectives.

1. Technical Basis for Management Measures

Watershed management implies an underlying causal theory: given the objective of protecting water supplies and the assignment of certain rights and responsibilities to various implementing agencies, the target groups will behave in the prescribed fashion and drinking water supplies will be protected. At the outset of a watershed management program, it may be possible for water systems to gain political support for their actions by basing drinking water protection efforts on the general theory that unmanaged watershed land uses will endanger the future adequacy and quality of water supplies. Before long, however, an effective management program will begin imposing costs on taxpayers and regulated landowners. To the extent that the costs cannot be justified by measurable and predictable improvements in the problems being addressed, political support for management programs will decline.[34]

The ability of various management measures to mitigate watershed problems depends upon two factors: (1) the ability of the measure to produce changes in target group behavior (e.g., to induce them to adopt various practices to reduce the generation and transport of pollutants); and (2) the ability of the practices, once adopted by the target groups, actually to affect the problem. In referring to the technical basis for watershed management measures, we are focusing on the latter phase of this two-step process of cause and effect. (The first step is also important and will be discussed in the following section.) In order to gauge the certainty with which prescribed practices will lead to improvements in water quality, analysts should utilize the information generated in the problem analysis stage of the planning process described in Chapter III.

2. *Target Group Behaviors to Be Changed by Management Measures*

Nonstructural management measures are designed to achieve desired watershed practices by changing target group behavior. A number of factors have been found to affect the extent of target group compliance with regulations and other measures designed to change their behavior. These factors include the proportion of the population affected by the management measure, the diversity of target group behaviors which are to be changed, the amount of change in behavior that is required, and the equity of the proposed changes in behavior (distribution of costs and benefits). Each of these factors should be weighed in estimating the effect of target group behavior on the feasibility of the alternative management measures being evaluated.

The proportion of the population whose behavior needs to be changed to protect drinking water supplies will affect the ease with which various management measures are implemented. In most instances, the smaller and more definable and isolated a target group, the more likely that political support can be mobilized and program objectives achieved. This would suggest that as a general rule watershed management measures should be initiated before a watershed is heavily developed and there are more landowners and other interests who might object to land use restrictions. In addition, the size of the target population should be considered when evaluating and comparing particular management measures. For example, in order to reduce sedimentation pollution, a performance-oriented zoning ordinance or a sedimentation pollution control ordinance could be used. Zoning would affect all of the property owners in the area, while a sedimentation pollution control ordinance would affect only land developers and those engaging in land disturbing activities. Because it affected fewer persons, the sedimentation pollution control ordinance would score higher in terms of political feasibility.

In general, the more diverse the behaviors that need to be changed by a watershed management measure, the more difficult it becomes to formulate precise voluntary or mandatory water quality protection rules and standards and thus the less likely that program objectives will be attained. For example, one major obstacle confronting the implementation of watershed ordinances designed to control the use of potentially hazardous chemicals is the extreme diversity in

the type, seriousness, mode of production, and use of the 325
organic chemicals that have been identified in drinking
water. The extent to which these pollutants are a problem
depends upon the industrial process used by firms which might
locate in a watershed, methods used to store and transport
chemicals, and natural conditions within a watershed. The
variety of factors to be considered means that precise
standards for water quality protection are difficult, if not
impossible, to devise. On the other hand, if water quality
protection is left to the discretion of field personnel,
program effectiveness may suffer because they may vary in
their technical competence and commitment to program
objectives.[35/36/37]

The extent of behavioral change required to achieve
watershed management objectives depends on the number of
people targeted by a program and the amount of change
required of each. It is obvious that the greater the amount
of behavioral change, the less likely that a management
measure can be easily implemented. Moreover, it is probably
easier to modify the behavior of a few large institutions
than it is to change the behavior of a great many individuals.[38] For example, one approach to watershed management
is a capital improvements program that avoids placing growth
inducing public services such as public sewerage and water in
water supply watersheds. This management program, which
concentrates on the policy decisions of one jurisdiction, is
more likely to be effectively implemented than one that
requires a large number of individual landowners to properly
install and operate on-site water quality protection devices.
In short, it is easier to monitor a few institutions for
compliance than many individuals.

The degree to which watershed management objectives are
attained will also be affected by how the costs of management
measures are distributed among target groups. Management
initiatives are less likely to succeed when some individuals
are asked to pay a disproportionate share of the costs
imposed by the program. For example, poor people are likely
to face greater economic hardships from costs imposed by
development regulations than are more wealthy individuals.
Moreover, the people living within a watershed are not apt to
be very pleased by the prospect of bearing most of a
program's costs, such as having to comply with development
restrictions, while the benefits of watershed management
accrue to the community at large.

Variation in the ability of target groups to pay the
costs imposed by watershed management measures will produce
pressures for flexible regulations and administrative

discretion. As noted above, such discretion increases the
chances that program objectives will not be achieved.[39]
Frequently, however, uniform standards are not a better
solution to the problem of diversity in target groups, since
they increase opposition from those who must bear a
disproportionate share of the costs.[40/41] In either case, it
will be more difficult to achieve program objectives when
watershed management must address diverse socioeconomic
groups.

3. *External Support for Management Measures*

If watershed management programs are to succeed, they
need support from a variety of groups external to the water
supply agency. These groups include the media, the public at
large, constituency groups, and local elected officials. In
evaluating the feasibility of alternative management
measures, each measure should be considered in terms of its
potential for gaining and maintaining support from these
groups.

Support provided for watershed management programs by
television, radio and newspapers will have a large bearing on
the amount of public support for water supply protection
efforts. The linkage between media attention on watershed
management and public support exists for two reasons. First,
the press is generally crucial for developing public
understanding of the importance of watershed management
efforts. Second, the tendency for the press to focus on an
issue for a short amount of time, and then to go on to
something else, is a real obstacle to obtaining a constant
infusion of political support for management activities. The
potential for using the press to acquire public support for
program objectives must be carefully considered in choosing
among alternative management measures. Moreover, once
management efforts are initiated, constant attention is
needed to maintain press interest in watershed management
throughout the course of the program.

The diffuse public support brought about, in part,
because of sympathetic media attention to watershed problems
needs to be converted to specific support from the organizations (constituency groups) that will be affected by the
management program. Local experience illustrates the value
of constituency groups. In Baltimore County, Maryland, for
example, farmers and their Farm Bureau strongly supported
growth management efforts because one objective of the
county's program, in addition to protecting water supplies,
was farmland preservation. This support has enabled the

county to withstand intense pressure from developers who continue to fight for a relaxation of growth management controls.[42]

As suggested by this example, potential watershed management measures should be considered in terms of their ability to garner support from various groups with objectives that go beyond high quality drinking water. At the same time, some caution is also in order. There are questions, for example, about the appropriateness of advocating farmland preservation as part of a watershed management program, since agriculture is a major polluter of many water supplies (see Chapter II). In addition, there is always the danger that secondary objectives of watershed management, such as increased recreational opportunities, will become a major concern and will usurp support for protecting water supplies.

Finally, watershed management measures have to be evaluated in terms of their acceptability to local elected officials. The amount of support local officials will give to watershed management will depend on: (1) the attention that the press gives to the issue; (2) public support for watershed management; (3) the existence of constituency groups to maintain interest in watershed management; and (4) the fiscal and economic stakes that a community has in protecting public water supplies over which it has jurisdiction.

The level of backing local officials give watershed management efforts should be considered in choosing a course of action to be taken by a water system in protecting its drinking water sources. In a situation where elected officials appear to care little about protecting drinking water sources, it may be necessary to concentrate on educational activities, direct subsidies for water quality protection to landowners, or to ask the state for legislative or regulatory actions to address water supply protection needs. It is important not to alienate public officials because their eventual support for watershed management is essential if program objectives are to be achieved.

4. *Internal Support for Watershed Management Measures*

Finally, management measures should be considered in terms of the level of support they can muster within the agencies that will administer them, including the water system itself. If agency personnel are not convinced of the value of a watershed management measure and are not committed to its implementation, it is unlikely that it will be applied

223

effectively. Two factors are likely to affect internal support for watershed management measures. First, the priority that an agency places on protecting public water supplies is important. While a water system may rank source protection very highly, other governmental agencies may be equally or more concerned with a variety of other planning functions and management issues, such as economic development, transportation, or recreational opportunities. Obtaining agency personnel's commitment to protect water supplies will be more difficult when their attention is divided by other social and economic problems, or when watershed management would drain resources from other agency programs. On the other hand, as noted above, some watershed management measures may contribute to the accomplishment of a number of goals in addition to source protection. In evaluating the political feasibility of measures, attention should be given, as noted above and in Chapter IV, to their ability to contribute to other goals of the water system and cooperating agencies.

A second factor which will affect the level of agency commitment to particular watershed management measures is availability of the professional and technical personnel needed to administer the measure. Obtaining support from personnel within the water system and from those associated with other agencies will be difficult if the expertise needed to understand and administer the measures is not currently available. While this constraint may be overcome by the addition of new program personnel with the requisite skills, the added administrative costs create another barrier to program adoption.

Summary

In addition to choosing a course of action which is cost effective and legally defensible, watershed planners and managers must select management measures which are politically feasible—measures which will be adopted and administered effectively by the water system and cooperating agencies. Estimating political feasibility requires subjective judgments about a number of factors which make measures more or less feasible. Here we have emphasized four key considerations: (1) the technical basis for the measure; (2) the target group behaviors the measure seeks to change; (3) external support from the media, potential constituency groups, and public officials; and (4) internal support from water system personnel and the personnel of cooperating agencies. The relative importance of each of these factors is likely to vary from one situation to another. In general, however, watershed management measures should pass muster on each of them, since failure to do so can result in the failure of the management program.

IMPLEMENTING A WATERSHED MANAGEMENT PROGRAM

Finally, after a community has established a capacity to undertake watershed management and has designed a program, it is ready to implement water supply protection measures. Implementation consists of three important steps which include:

1. Developing specific policy, program measures, and substantive and procedural directives within the intervention measures.

2. Incorporating appropriate sanctions and inducements to assure compliance with management initiatives.

3. Maintaining sustained public involvement in watershed management actions.

Another step which could be included above is program evaluation and revision. However, it is instead the topic of the entire next chapter because of its critical importance to successful management of watersheds.

Specific Policy and Directives

Even after management tools have been selected, policies and intervention measures must be drafted in specific language that provides clear directions to the target groups and to the local officials who will administer the measures. The target groups need to know how to comply with regulations or to take advantage of incentives. Local officials must know how to administer the devices. Program implementation is simplified and made more effective when policies and regulations are specific and clearly written.

In addition to providing a clear idea about legislative intent, standards and procedures, specific directives are important for several other reasons. First, they make the job of explaining the program to the public much easier since its rules are not obscured by vague policy or legal jargon. Second, the program is simpler to evaluate if actual standards and resulting ambient qualities that it is supposed to achieve are precisely stated. Third, clarity helps indi-

225

viduals plan investments within watersheds because they know
at the outset what they are able to do with their land.
Finally, specific policies and regulations make it possible
for concerned citizens and iterest groups to monitor
implementation.

Sanctions and Inducements

It is a simple fact that laws will be ineffective unless
there are adequate sanctions if people fail to comply with
them. Many watershed regulations impose high costs on land
users. If penalties for violating these laws are not
stringent enough, then people will break them in order to
save money.

On the other hand, regulations may be self defeating
because they are too strict. While the initial support for
enacting watershed management measures almost always subsides
with time, resistance to regulations will grow as the costs
they impose are felt. People fighting watershed protection
measures will become better organized and politically
powerful, and eventually may be successful in having them
repealed. Thus, it is important to consider the use of
incentives to persuade land users to protect drinking water
supplies.

Incentives may take many forms. Focusing attention on
threats to public health caused by polluting activities may
be enough to persuade some individuals to take precautions to
protect water quality. Direct cash payments or tax breaks to
defer the costs of antipollution practices are another type
of incentives. Whatever form they take, incentives are
useful because they can spread the costs of watershed
protection measures to everyone who benefits from them, not
just the people who are regulated.

In addition to supplementing regulations, incentives are
important ways to combat those pollution problems that cannot
be solved by mandatory requirements. Cost-sharing programs
that provide funds to install soil conservation practices on
farmland are about the only way agricultural pollution
problems can be addressed since few communities have been
willing to enact regulations pertaining to farming.
Political resistance to regulating any type of land use may
also necessitate the use of incentives to protect water
supplies.

Sustained Public Involvement

Although initial efforts have already been made to
develop public awareness and support for watershed
management, it is vital to continue to involve the public in
protecting water supplies. As discussed previously, even
strong support for watershed management at the outset of the
program may be eroded as the press turns its attention
elsewhere and costs of the program are felt. Therefore,
aggressive steps are needed first to obtain public support
and then to maintain it over the course of the program.

Prior to adoption of watershed management measures or at
the beginning of a renewed management effort, public support
for watershed management can be promoted through educational
activities. A variety of measures can be employed to educate
the public about why watershed management is important.
Public meetings and hearings, newspaper articles and letters
as well as public service announcements on radio and television are just a few examples of tools available for public
education. There are also direct ways of reaching water
system customers and watershed landowners that should be considered as part of an educational program. Brochures about
watershed management needs included with water bills is one
direct approach. Another is to conduct sanitary surveys in
watersheds to make landowners aware of the water quality
problems they are causing by poor soil stewardship or other
hazardous land uses. Moreover, the findings of such inspections can be reported to the press to focus public attention
on watershed conditions. Whatever methods are used to obtain
public support for watershed management, the overriding concern is to acquire enough understanding about watershed
management requirements so that the public will actively
support and become involved in watershed management initiatives. Without this citizen support, it will usually be
impossible to muster enough political support to enact watershed management programs.

Continued public involvement requires more than one-time
educational events. Public participation in watershed
management activities should begin at the outset of a program
by designating active citizen advisory committees or other
similar task forces to look at watershed management
requirements.* News reporters should be asked to serve on
these committees to develop a deeper understanding and con-

---

*In the City of Charlottesville and Albemarle County,
Virginia, citizen task forces were instrumental in
establishing watershed management policies and obtaining the

tinued coverage of watershed management in the local press. Moreover, well publicized media events can routinely be staged, such as fishing contests or boat races in the water supply reservoir, to focus public attention on watershed management activities.

Two other techniques that help to sustain public involvement are public notices of proposed development projects in watersheds and local ordinances that reqire environmental impact statements for any new watershed land use. Both of these measures provide citizens with the opportunity to monitor and become active in decisions about how water supply watersheds should develop.

SUMMARY

The design and implementation of watershed management programs is a complicated process that demands knowledge, skill, and common sense to be carried out successfully. Building a team which embraces these skills--scientific understanding of natural processes in the watershed, engineering knowledge of physical control techniques, land use planning, understanding of various intervention measures, policy/economic analysis expertise in cost effectiveness evaluation, legal knowledge of institutional alternatives and constitutional standards, political skills needed to adopt and implement a watershed management program, and understanding of techniques for involving citizens in the management process--is the first order of business in program design.

Once watershed problems have been analyzed, program targets have been set and potential management measures have been identified, a course of action must be selected. Three sets of analyses for choosing among measures have been discussed in this chapter. Cost effectiveness evaluations indicate which measures will achieve program objectives at what costs, both monetary and non-monetary. Institutional evaluations suggest which agencies should be looked to to adopt and implement particular measures and how those measures should be framed to withstand possible legal challenges. Finally, political feasibility evaluations suggest the likelihood that various management measures being considered will be adopted by various agencies and remain viable over the period of the program. Taking all of these

---

necessary political support for local governments to act to protect public water supplies.[43]

factors into account in formulating a watershed management program is more art than science. Nevertheless, in this chapter we have discussed a number of considerations which should improve the chances of making the right choices.

The chapter concluded by noting that the chances for successful program implementation can be improved substantially if program implementation incorporates three additional steps. These include developing specific policy and regulatory derectives, incorporating appropriate sanctions and inducements to improve compliance with management directives and regulatory requirements, and maintaining sustained public involvement.

REFERENCES

1. Miller, Todd L. and Raymond J. Burby with Edward J. Kaiser and David H. Moreau. Protecting Drinking Water Supplies Through Watershed Management: A Casebook for Devising Local Programs, Center for Urban and Regional Studies, The University of North Carolina at Chapel Hill, Chapel Hill, N.C., August 1981.

2. Sabatier, Paul and Daniel Mazmanian. "Toward a More Adequate Conceptualization of the Implementation Process—With Special Reference to Regulatory Policy," Public Policy, July 1978.

3. Guidelines for Areawide Waste Treatment Management Planning, U.S. Environmental Protection Agency, Washington, D.C., 1975.

4. Interagency Committee on Water Resources, Subcommittee on Evaluation Standards, Proposed Practices for Economic Analysis of River Basin Projects, U.S. Government Printing Office, Washington, D.C., 1950.

5. McKean, Roland N. Efficiency in Government Through Systems Analysis, John Wiley and Sons, New York, 1958.

6. Maass, Arthur, Maynard M. Hufschmidt, Robert Dorfman, Harold A. Thomas, Jr., Stephen A. Marglin, and Gordon Maskew Fair. Design of Water-Resource Systems, Harvard University Press, Cambridge, Mass., 1962.

7. Mishan, Edward J. Cost-Benefit Analysis, Praeger Publishers, New York, 1976.

8. U.S. Water Resources Council. "Establishment of Principles and Standards for Planning," _Federal Register_, Vol. 38, No. 174, 1973.

9. Haveman, R.H. _The Economic Performance of Public Investments_, Resources for the Future, Inc., Washington, D.C., 1972.

10. North, R.M., A.S. Johnson, and H.O. Hillestad with P.A.R. Maxwell and R.C. Parker. _Survey of Economic-Ecologic Impacts of Small Watershed Development_, Report No. ERC 0974, Institute of Natural Resources, The University of Georgia, Athens, in cooperation with Environmental Resources Center, Georgia Institute of Technology, June 1974.

11. Miller, Todd L. and Raymond J. Burby with Edward J. Kaiser and David H. Moreau. _Protecting Drinking Water Supplies Through Watershed Management: A Casebook for Devising Local Programs_, Center for Urban and Regional Studies, The University of North Carolina at Chapel Hill, Chapel Hill, N.C., August 1981.

12. _Evaluation of the Cost-Effectiveness of Nonstructural Pollution Controls: A Manual for Water Quality Planning,_ Water Planning Division, U.S. Environmental Protection Agency, April 30, 1976.

13. Miller, Todd L. and Raymond J. Burby with Edward J. Kaiser and David H. Moreau. _Protecting Drinking Water Supplies Through Watershed Management: A Casebook for Devising Local Programs_, Center for Urban and Regional Studies, The University of North Carolina at Chapel Hill, Chapel Hill, N.C., August 1981.

14. Eubank v. Richmond, 226 U.S. 137 (1912).

15. Lawton v. Steele, 152 U.S. 133 (1894).

16. Ira, Paul. "Constitutional Law: Land Use Regulation: You Don't Have to Take It or Leave It," _University of Florida Law Review_, Vol. 31, p. 431.

17. Uchtmann, D.L. and W.D. Seitz. "Options for Controlling Non-Point Source Pollution: A Legal Perspective," _National Resources Journal_, Vol. 19 (1970), p. 600.

18. Rathkopf, Arden H. _The Law of Zoning and Planning_, Clark Boardman Co., Ltd., New York, 1975, Vol 1, §8-5,

19. Uchtmann, D.L. and W.D. Seitz. "Options for Controlling Non-Point Source Pollution: A Legal Perspective," National Resources Journal, Vol. 19 (1970), p. 600.

20. Ira, Paul. "Constitutional Law: Land Use Regulation: You Don't Have to Take It or Leave It," University of Florida Law Review, Vol. 31, p. 429.

21. 438 U.S. 104 (1979).

22. 369 U.S. 590 (1961).

23. "State and Local Wetlands Regulation: The Problem of Taking Without Just Compensation," Virginia Law Review, Vol. 58 (1972), p. 876.

24. Enterprise Irrigation District v. Farmers' Mutual Canal Company, 243 U.S. 157, at 164 (1916).

25. Miller, Todd L. and Raymond J. Burby with Edward J. Kaiser and David H. Moreau. Protecting Drinking Water Supplies Through Watershed Management: A Casebook for Devising Local Programs, Center for Urban and Regional Studies, The University of North Carolina at Chapel Hill, Chapel Hill, N.C., August 1981.

26. McQuillin, Municipal Corporations, §15.30.

27. Platt, Rutherford H., George M. McMullen, Richard Paton, Ann Patton, Michael Grahek, Mary Read English, and Jon A. Kusler. Intergovernmental Management of Floodplains. Institute of Behavioral Science, University of Colorado, Boulder, Co., 1980.

28. Lynn L. Schloesser. "Agricultural Nonpoint Source Water Pollution Control under Sections 208 and 303 of the Clean Water Act: Has Forty Years of Experience Taught Us Anything?" North Dakota Law Review, Vol. 54 (1978), p. 605.

29. Platt, Rutherford H., George M. McMullen, Richard Paton, Ann Patton, Michael Grahek, Mary Read English, and Jon A. Kusler. Intergovernmental Management of Floodplains. Institute of Behavioral Science, University of Colorado, Boulder, Co., 1980.

30. Miller, Todd L. and Raymond J. Burby with Edward J. Kaiser and David H. Moreau. Protecting Drinking Water Supplies Through Watershed Management: A Casebook for Devising Local Programs, Center for Urban and Regional Studies, The University of North Carolina at Chapel Hill, Chapel Hill, N.C., August 1981.

31. 129 N.E. 2d at 685, citing Ill. Rev. Stat., 1953, Chap. 24, 75-3.

32. Miller, Todd L. and Raymond J. Burby with Edward J. Kaiser and David H. Moreau. Protecting Drinking Water Supplies Through Watershed Management: A Casebook for Devising Local Programs, Center for Urban and Regional Studies, The University of North Carolina at Chapel Hill, Chapel Hill, N.C., August 1981.

33. Miller, Todd L. and Raymond J. Burby with Edward J. Kaiser and David H. Moreau. Protecting Drinking Water Supplies Through Watershed Management: A Casebook for Devising Local Programs, Center for Urban and Regional Studies, The University of North Carolina at Chapel Hill, Chapel Hill, N.C., August 1981.

34. Sabatier, Paul and Daniel Mazmanian. "Toward a More Adequate Conceptualization of the Implementation Process—With Special Reference to Regulatory Policy," Public Policy, Vol. 6 (July 1978).

35. Sabatier, Paul and Daniel Mazmanian. "Toward a More Adequate Conceptualization of the Implementation Process—With Special Reference to Regulatory Policy," Public Policy, Vol. 6 (July 1978).

36. Schultze, Charles. The Public Use of Private Interests, Brookings Institution, Washington, D.C., 1977.

37. Elmore, Richard. "Organization Models of Social Program Implementation," Public Policy, Vol. 6 (Spring 1978).

38. Sabatier, Paul and Daniel Mazmanian. "Toward a More Adequate Conceptualization of the Implementation Process—With Special Reference to Regulatory Policy," Public Policy, Vol. 6 (July 1978).

39. Sabatier, Paul and Daniel Mazmanian. "Toward a More Adequate Conceptualization of the Implementation Process—With Special Reference to Regulatory Policy," Public Policy, Vol. 6 (July 1978).

40. Sabatier, Paul and Daniel Mazmanian. "Toward a More Adequate Conceptualization of the Implementation Process--With Special Reference to Regulatory Policy," Public Policy, Vol. 6 (July 1978).

41. Lieber, Harvey. Federalism and Clean Waters, Heath Publishers, Inc., Lexington, Mass., 1974.

42. Miller, Todd L. and Raymond J. Burby with Edward J. Kaiser and David H. Moreau. Protecting Drinking Water Supplies Through Watershed Management: A Casebook for Devising Local Programs, Center for Urban and Regional Studies, The University of North Carolina at Chapel Hill, Chapel Hill, N.C., August 1981.

43. Miller, Todd L. and Raymond J. Burby with Edward J. Kaiser and David H. Moreau. Protecting Drinking Water Supplies Through Watershed Management: A Casebook for Devising Local Programs, Center for Urban and Regional Studies, The University of North Carolina at Chapel Hill, Chapel Hill, N.C., August 1981.

CHAPTER VII

PROGRAM MONITORING AND EVALUATION

It is estimated conservatively that the United States government alone spends more than a half-billion dollars for evaluations every year.[1] Whatever the program being evaluated, it is critical for policy makers, funding organizations, planners, and program staffs to have answers to four basic questions:

1. Is the intervention reaching the appropriate target groups?

2. Is it being implemented in the ways specified?

3. Is it effective?

4. What are its costs relative to its effectiveness?

Answers to these questions are not always easy to obtain, especially when it comes to managing a water supply watershed. Yet it is important for water systems to pose these questions, as is illustrated by the following four examples:[2]

1. Norfolk, Virginia, spent $120,000 to conduct an assessment of watershed management needs for its water supply reservoirs. The final report of this investigation has never been seen by the planning staffs of the jurisdictions that actually control land uses within the watersheds. An evaluation would be useful to determine why the effort did not reach the people who need it and what should be done to correct this situation.

2. Albemarle County, Virginia, requires stormwater detention devices for new development. In one instance, raised curbs and gutters designed to slow runoff off a parking lot were placed uphill from the flow of water. An evaluation would be useful to determine if this is an

isolated occurrence or if other program
requirements also are not being properly
implemented, operated, and maintained.

3. Newark, New Jersey, has devised a comprehensive
    plan to develop commercial, residential, and
    recreational land uses on the 30,000-acre water
    supply watershed that it owns. Even though the
    plan has withstood court challenges questioning
    its environmental sensitivity, some water
    quality experts still contend that if it is
    implemented, drinking water quality will be
    degraded. As the plan is realized, evaluations
    are needed to assess its effectiveness in
    protecting drinking water supplies.

4. Baltimore County, Maryland, has been engaged in
    an $800,000 planning program which is aimed
    partially at protecting three water supply
    reservoirs from pollution. Large lot zoning
    and strict stormwater detention requirements
    are some of the controls that have been
    implemented to protect drinking water. One
    developer has challenged the program in court,
    saying that the water quality benefits received
    do not justify the increased development costs
    imposed by the program. An evaluation that
    determines development program costs relative
    to the value of water quality improvement would
    be useful evidence to settle this debate.

These examples serve to illustrate the important role
that evaluations can play in watershed management. If
designed properly, an evaluation will not only assist in
overcoming problems that interfere with effective watershed
management, but can also help to avoid new problems in the
future. However, utmost care should be taken in conducting
program assessments, for if they are not well designed they
may actually detract from effective management efforts.

Most of the evaluation procedures discussed in this
chapter require some degree of estimation and interpolation.
This is particularly true with the impact assessment and
cost-effectiveness analysis methods that are covered, since
in using each method hypothetical programs must be constructed. However, these evaluation techniques are
applicable to any stage in the watershed management process--
during program design, ratification, or after implementation.

236

GUIDELINES FOR DESIGNING WATERSHED MANAGEMENT EVALUATIONS

Seven basic guidelines that are important to consider when designing an evaluation of watershed management activities are:

1. Evaluations should be viewed as an integral part of the overall watershed planning and management effort.

2. Evaluations should involve all government agencies that are contributing to the watershed management effort.

3. Private individuals affected by watershed management activities should be participants in an evaluation.

4. Evaluations should yield results that are positive and constructive so that they are rewarding and unthreatening to program participants.

5. Particular care should be taken when deciding what data must be collected to evaluate watershed management activities.

6. Data collection efforts should be institution-alized into management activities.

7. Designers of watershed management evaluations must recognize the limitations of such an assessment.

First, evaluations should be viewed as an integral part of the overall watershed planning and management effort. Evaluations are frequently treated as discrete activities, functionally separate from other program activities, with those responsible for undertaking evaluations having little or no influence over the nature of preceding work. Moreover, evaluations are often left until too late in the planning process for them to make an effective contribution to subsequent decisions. Under these circumstances it is likely that data necessary to measure the consequences of a program will not have been collected, or the data that are available will be in a form that makes them difficult to use.[3] Thus, if evaluations of watershed management are to be useful, they should actually begin to be undertaken at the outset of the management program.

Second, it is important that an evaluation involve all the government agencies that are contributing to the watershed management effort. Local governments and regional and state agencies have personnel who not only have considerable interest in the evaluation, but who also can do the most with its findings. Unfortunately, it is not always easy to gain the cooperation of these people because of lack of staff time, interest, or training in evaluation methods.[4]

Third, in addition to involving government agencies, private individuals affected by watershed management activities should be participants in an evaluation. Interested citizens, including watershed landowners and members of the land development community, can provide valuable information on how well the program is being administered and received, as well as its effect on the use of watershed property.[5]

Fourth, evaluations should yield results that are positive and constructive so that they are rewarding and unthreatening to program participants. This may be achieved partially by including all interested parties early in the evaluation process so that they fully understand its intentions. In addition, attention should be focused on identifying solutions to problems highlighted by an evaluation, rather than just to attributing blame for program shortcomings. Finally, an effort should continually be made to congratulate those agencies and people for a job well done.[6]

Fifth, particular care should be taken when deciding what data must be collected to evaluate watershed management activities. Eight criteria for selecting evaluation data include: (1) appropriateness and validity; (2) uniqueness; (3) completeness; (4) comprehensibility; (5) controllability; (6) cost; (7) timeliness of feedback; (8) accuracy and reliability.[7] Each of these criteria is described in Table VII-1.

Sixth, data collection efforts should be institutionalized into management activities. This means that building permits and other administrative records should be formalized and made a part of the annual budget-planning process. If data collection is not a routine exercise, the process will remain vulnerable to budget cuts as personnel come and go and as elected officials criticize government expenditures.[8]

Seventh, designers of watershed management evaluations must recognize the limitations of such assessments. For example, even though the water system may have the best of

Table VII-1. Criteria for Selecting Evaluation Data

Criteria	Explanation
Appropriateness and Validity	Does the measure relate to watershed management objectives and does it really measure the degree to which community needs are being met, including minimization of detrimental effects?
Uniqueness	Does it measure some effectiveness characteristic that no other measure encompasses?
Completeness	Does the list of measures cover all or at least most of the program impacts?
Comprehensibility	Is the measure understandable?
Controllability	Does the condition measured at least partially result from watershed management? Does the government have some control over it?
Cost	Are cost and staffing requirements for data collection reasonable?
Timeliness of Feedback	If the information is needed for specific decisions, possible launching of a new program, setting budget levels for the coming year, and the like, wit the data and analysis become available before the decision makers reach their deadline?
Accuracy and Reliability	Can sufficiently accurate and reliable information be obtained? This is a problem not only with procedures that use samples, such as citizen surveys, but with many government statistics, such as water quality indicators.

Adapted from: Harry P. Hatry, Louis H. Blair, Donald M. Fisk, John M. Griener, John R. Hall, Jr., and Philip S. Schaenman, How Effective Are Your Community Services: Procedures for Monitoring the Effectiveness of Municipal Services, The Urban Institute, Washington, D.C., 1977.

intentions when it evaluates watershed management measures administered by local governments which control activities taking place on its watersheds, by "meddling" in other governments' affairs it is likely to encounter political opposition and institutional jealousies. Another limitation stems from the lack of scientific understanding about land use and water quality relationships which makes it more difficult to measure accurately the effects of the watershed management activities. Evaluations must assess indicators of program effectiveness rather than attempt to make precise measurements.[9]

TYPES OF WATERSHED MANAGEMENT EVALUATIONS

At the beginning of this chapter four basic questions that watershed management program evaluations should seek to answer were listed. Three types of evaluations are commonly conducted to address these questions: (1) program monitoring; (2) impact assessment; and (3) cost-effectiveness analysis.

First, program monitoring evaluations examine if the program is being administered properly and if target groups are complying as expected. This is normally a low cost type of evaluation that is conducted with interviews and compliance checks using public records. Program monitoring evaluations should be conducted at least on an annual basis to assure consistent implementation.

Second, impact assessments are conducted to identify the intended and unintended effects of a program and to attempt to determine if these impacts are a result of the program or some alternative process that does not include watershed management activities. There are a wide variety of impacts that might be associated with watershed management programs. Evaluations that attempt to assess all possible impacts frequently cannot be conducted because they are too costly, and thus communities are forced to limit the scope of their evaluations by how much they can afford. Impact assessments should be conducted only when the program has had enough time to cause impacts that are large enough to be measured.

Finally, cost-effectiveness analysis, described in the previous chapter, is useful to reassess if the costs (both private and public) to deliver watershed management benefits are worth it when compared with alternative uses of the same resources. The amount of money that a community spends on cost-effectiveness evaluations should be governed by the

severity of shortcomings in watershed management efforts
revealed by an impact assessment.

PROGRAM MONITORING EVALUATIONS

Monitoring of a watershed management program is conducted to determine whether specified practices and intervention efforts are being properly administered and whether
specified target groups are complying with program regulations and other requirements. It is important to assess
government programs along these lines for two reasons.
First, the credibility of the program with elected officials
and the general public is enhanced by showing that what
presumably was paid for and deemed desirable was actually
undertaken. Second, it is impossible to evaluate the impact
of a program unless it did indeed take place and target
groups actually did comply with its requirements.*

The first step that must be taken to conduct a program
monitoring evaluation is to clearly define the watershed
management effort. All phases of the program should be
described, including staff training and public education
elements as well as voluntary and mandatory land use
controls. In addition to program initiatives, the program
staff--including administrators and technicians--and the
resources it uses should be determined. Although seemingly
simple, this task may become very complex if the program is
administered by a variety of government agencies and
administrative duties are incorporated into these agencies'
programs. However, unless the management program is
explicitly defined, it will be difficult to know what to
measure.

Once the components of the management program that will
be evaluated are identified, the next task is to determine
who the people are that are the target groups of management
initiatives. This list will not only include inhabitants of
the watershed, but also financial institutions, developers,
government agencies, and anyone else who has influenced the
development and use of watershed property. Moreover, it is
vital to consider prospective target groups that may be
expected to respond to program initiatives in the future.

---

*Rossi et al. devote an entire chapter to a discussion
of program monitoring techniques. The program monitoring
methods for watershed management discussed in this chapter
have been adapted from Rossi.[10]

The next step is to attempt to monitor program implementation. Implementation includes the extent to which target groups comply with program requirements, target group understanding of program goals and objectives, and an assessment of the quality of program administration. Finally, it is important to obtain suggestions from target groups about possible improvements in watershed management efforts.

There are four data sources that should be considered in designing a monitoring evaluation: (1) direct observation by the evaluator; (2) program records; (3) data from program administrators; and (4) information from program participants or their associates. It is unlikely that any one of these data sources will provide enough information alone and so all may have to be considered in conducting an evaluation.

For some purposes, the preferable and relatively inexpensive data collection approach for monitoring watershed management programs is direct observation. Direct observation by the evaluator is appropriate whenever the presence of an observer does not affect the manner in which the program is implemented. However, evaluators must be trained to assess conditions in a consistent and accurate manner. In addition, the observation effort must be systematic. Typically, three ways have been tried to make systematic observations. The first approach is simply a recording of events and the order in which they occur in as much detail as possible. Another observation method is to establish a set of questions that the evaluator must answer. A third approach is a structured rating scheme, such as checklists or multiple answer questions that avoid relying on the ability of the evaluator to describe what is observed.

Program records are a second source of data that provide information on implementation. Zoning variances, land development permits, property tax maps and cards all provide a wealth of information about program implementation. Three rules should be followed when relying on program records for data. First, a few items of data gathered consistently and reliably are much better for monitoring purposes than a more comprehensive set of information of doubtful reliability and inconsistent collection. Second, whenever possible, it is useful to prestructure record forms as checklists so that program staff does not have to provide narrative information. Third, it is important to review completed records for consistency and accuracy as carefully and as soon as possible to catch ommissions and inconsistencies.

Program administrators are a third important source of information for monitoring the implementation of watershed management programs. Staff members may be interviewed or asked to complete a survey that inquires about different aspects of program implementation. One problem with relying upon project personnel for information is that they may intentionally or unintentionally bias their accounts. However, the value of interviewing and surveying program staff should not be underestimated, especially since such methods reduce the time and effort that must be spent on evaluations.

The fourth source of data is the targets of the watershed management program. The information that groups such as property owners and developers can generate is valuable because: (1) watershed managers may not be aware of what is important to participants; (2) citizen reaction to program initiatives is necessary to orchestrate future implementation procedures; (3) it may be the only way to determine if the program is being properly administered in the field; and (4) it helps to find out if participants fully understand the purposes of watershed management activities. This information may be collected using surveys or interviews.

Once data have been collected, the final step in conducting a program monitoring evaluation is to determine if watershed management activities are being properly administered. An important question is the extent to which the program as implemented resembles the program as designed. This question can only be answered by examining for bias and competence in program administration and by estimating the actual coverage of the program and the responses of participants to management activities. Discrepancies between the watershed management program design and its implementation may result in respecification of the program or in improved administrative procedures. Program monitoring information also provides an opportunity to judge the appropriateness of conducting an impact evaluation or cost-effectiveness analysis. If a program has not been administered properly or as intended, it may make little sense to expend time and resources on measuring program impacts.

IMPACT ASSESSMENT

Impact assessments are directed at establishing, with as much certainty as possible, whether or not an intervention is producing its intended results.* In order to do so, it is

---

*The development of the impact assessment approach can be traced to the cost-revenue studies of the 1950s. For

necessary to measure as rigorously as possible the outcomes
of a management program and to purify these results by
removing the influences of forces other than the program
being evaluated. The critical issue in impact evaluation is
whether or not a program has produced more of an effect than
would have occurred "naturally." Thus, to evaluate a watershed management program that attempts to reduce water pollution requires assessing whether or not water quality is
better than would have occurred if the program had not been
introduced. For instance, land development activity and
associated erosion and stream sedimentation may decline due
to housing market conditions, such as high interest rates,
which have no connection with the watershed management
program. Thus, it would be a mistake to attribute improved
water quality to the program. Failure to accurately assess
program impacts can result in a false sense of security about
how well water quality is protected or may lead to inappropriate modifications of various management activities.

Identifying Program Impacts

A key requirement for conducting impact assessments is
to know what impacts watershed management programs can have,
both intended and unintended. From the perspective of the
water system, the most important impact to assess
is the effect of the program on achieving a protected supply
of good quality drinking water. Unfortunately, this effect
cannot be easily measured due to two major problems. First,
it is difficult to measure water quality.[19] Periodic
sampling, even at daily intervals, frequently misrepresents
water quality because it does not reflect the rapid changes

---

example, see reference 11. In these initial studies communities tried to determine if the costs associated with proposed
land use changes would be balanced by the revenues the
changes generated. The approach was broadened from only
fiscal concerns to include a wider range of economic concerns, such as jobs and income. With the passage of the
National Environmental Policy Act in 1969, environmental
impact assessment was required for federal actions having
significant effect on the environment. Social impact assessment methodologies were also developed and applied to large
transportation and other projects which were thought to have
potentially disruptive effects on communities and neighborhoods. The leading explication of the impact assessment
approach is provided by a series of works published by the
Urban Institute. This Series includes references 12 through
15. See also references 16 through 18.

that constantly occur. Furthermore, routine water quality analysis is incapable of detecting some of the more exotic chemicals that may be present in water supplies.

Another difficulty in assessing the impact of the program on raw drinking water quality is in proving that water quality changes are related to land use practices within the watershed. There is too little information on the linkages between particular land uses and water quality. Moreover, the fate of many water contaminants in the aquatic environment and their threat to human health is unknown. These uncertainties combined with the high degree of natural diversity between watersheds makes it impossible to measure precisely the level of water quality protection that has been achieved.

Because of the difficulties associated with measuring the impact of a program on protecting drinking water quality, it is necessary to establish proxy measures that can be used to indicate indirectly how well the management program is working. One proxy measure is the raw water quality samples that are taken daily at water intake and treatment plants. Turbidity samples, as well as other parameters that are measured, will indicate major changes in water quality.

A second proxy measure is sedimentation rates in a reservoir. As discussed in previous chapters, many water pollutants are transported by sediment which is generated by a variety of land disturbing activities. In Baltimore County, a sediment survey of Loch Raven Reservoir indicated that it had lost substantial capacity and placed pressure on local officials to institute tighter restrictions on development within the lake's watershed.[20] Unlike water quality monitoring, sedimentation surveys are fairly easy and inexpensive to conduct. However, over-reliance on such surveys as indicators of program success may result in other water quality problems remaining undetected.

In addition to assessing changes in the condition of water supply reservoirs, an impact evaluation should also monitor changes in land use practices within watersheds which are attributable to the management program. This includes changes in the percentage of industrial, commercial, residential, or undeveloped land. Sanitary surveys are another type of measure that can be taken on a yearly basis. These surveys are conducted to determine if any hazards to water quality—such as failing septic tanks, chemical dumps, and the like—exist in the watershed. Actual measurements of watershed land uses and management activities may vary substantially, from acreage amounts to ordinances. The basic premise behind these proxy measures is that watersheds undergoing rapid land use changes and which are unmanaged will not have protected water supplies. Many of the problem analysis techniques discussed in Chapter III may be applied in assessing water quality impacts of a management program.

There are a number of other dimensions besides land use and water quality along which impacts of watershed management programs can be divided. These impacts may be intended or unintended and should be considered secondary effects of the management program. Three types of impacts are commonly recognized as resulting from land use decisions: economic, social, and environmental. Although these are considered secondary impacts when compared with water quality, they are important to consider because their occurrence may have a bearing on the program's long-range success.

Economic impacts are critically important in determining community acceptance of watershed management activities. Management activities may alter the costs and revenues accruing to government. Important economic impacts include those on local governments, since they are responsible for approving watershed management activities. The most obvious fiscal impact is the operating cost of implementing the watershed management program itself. The labor and legal costs of programs with major regulatory and inspection components can be high, but they are negligible when compared to watershed management programs which entail significant land acquisition. Watershed management may, however, be able to offset at least a portion of these costs by reducing the costs of public services. Since an unprotected water supply is likely eventually to have to be replaced, watershed management may result in substantial savings. Moreover, if the program results in more compact urban development elsewhere in the community, expenditures for public water and sewerage may be substantially reduced. Savings on the costs of treating raw drinking water, which presumably will be of higher quality because of watershed management, should also be considered.

On the revenue side, negative fiscal impacts may be generated by the watershed management program. Any changes in property values will affect revenues generated through the property tax. Land left as open space or in agricultural and forest uses will not provide the level of taxes associated with urban uses. Similarly, watershed lands acquired for public uses will produce no tax revenue at all. Local governments can be expected to be wary of the fiscal impacts of watershed management programs.*

Economic effects of watershed management programs in the
community are another set of key impacts to be studied. By
providing a protected water supply, a watershed management
program helps to assure the continued economic viability of a
community. The degree to which watershed management affects
local economic development is difficult to determine. One
possible way to gauge this impact is to question investors
about the importance of having a good water supply in making
location decisions.

Some economic impacts of watershed management are often
major obstacles to the adoption and implementation of an
effective program. One of the most important is the presumed
negative impact on property values in the watershed. Recent
case studies, however, indicated that in some instances
management programs may actually increase the value of
watershed property by making it a more exclusive place to
live.[23] Moreover, economic theory suggests that a program
which severely restricts new watershed development would
increase the value of existing watershed development and
vacant land outside of the watershed while decreasing the
value of vacant watershed land. All these impacts and their
net effect may be important economic impacts of watershed
management.

Increased construction costs due to pollution control
requirements and larger lot sizes may increase the cost of
housing and other development, especially if a large portion
of the community is within a water supply watershed. Perhaps
the most difficult economic impacts to measure are those
which result from displacing development from the watershed.
Such development may remain within the community, but be
forced to choose a less (or more) efficient location or it
may leave the community entirely if alternative sites are not
available. These increased costs and lost jobs and income
are legitimate impacts to assess.

The environmental impacts of watershed management are
generally thought to be positive. It is useful to divide
environmental impacts into three types: (1) natural resource

---

*In 1979, the U.S. Department of Housing and Urban
Development published The Fiscal Impact Guidebook to acquaint
local officials with the various methods of estimating fiscal
impacts.[21] In 1980 it also published A Practitioner's Guide
to Fiscal Impact Analysis to assist in using its earlier
guidebook.[22]

conservation; (2) environmental quality, and (3) amenity. Natural resource conservation means the preservation of such resources as unique ecosystems, wetlands, and prime agricultural land. Environmental quality refers to the introduction of pollutants into the environment. Watershed management should generally have a positive impact on environmental quality since a prime objective of the program is to maintain safe drinking water supplies. Lastly, environmental impacts may take the form of effects of a community's amenities. The preservation of open space or agricultural land in the watershed may simply be valuable for recreational purposes or for visual relief from the urban setting. This preservation of beauty is an aesthetic judgment, but is a potential environmental impact of watershed management which should not be overlooked in evaluating a program's impacts.

Social impacts are probably the most diverse and the most elusive results of watershed management. The most important social impact is improved health because of drinking water. The elimination of health hazards due to polluted water supplies is a positive social impact of watershed management. Through acquisition and preservation of open space a management program can also enhance recreational opportunities in the community.

There are also potentially negative social impacts of watershed management. Aggressive acquisition programs and large-lot zoning may increase housing costs and exclude lower income groups from the watershed or the entire community. Watershed landowners may also resent what they view as undue restriction of their property rights.

The equity of the distribution of all types of impacts among groups may also be considered a social impact. As noted earlier, the impacts of a watershed management program do not affect all groups equally. The question of equity is concerned with who wins and who loses as the result of management policies. There are a number of different ways in which equity issues may be analyzed. Groups affected by the watershed management program may be defined by income level, since governments generally wish to avoid placing the heaviest burdens on those least able to bear them. They may be defined spatially, since it may be viewed as inequitable if program beneficiaries are forced to live outside the watershed while those bearing the costs live within it. Groups may also be defined intergovernmentally, since there is a growing consensus that higher levels of government should not be forced to bear the costs of measures which benefit only specified local areas.

No matter which of these distinctions is used, the basic equity concept requires that those who benefit from watershed management should share in its costs. The law guarantees a minimum level of equity by requiring just compensation for property taken for public use and by imposing other limitations on the use of local government police powers. The equity of watershed management will depend on the correspondence of its costs and benefits among affected groups.

Strictly speaking, equity is usually not considered an impact in and of itself. It is rather the distribution between groups of net gains and losses resulting from the impacts discussed in the previous sections. However, the best way to include equity within an evaluation is to consider it a social impact. Thus, a program which distributes the other impacts inequitably would be seen as producing negative social impacts.

While all impacts may not be present in every community or may not be important in light of the objectives used in an evaluation, they should at least be recognized as potential results of watershed management. Prior to measurement, a preliminary screening should be conducted to determine which impacts are likely to be present in a particular community and relevant to the objectives of the evaluation.

Measuring Impacts

Accurate measurement of impacts is central to the whole process of evaluating watershed mangement. One problem that will be encountered is that each set of impacts (economic, fiscal, social, and environmental) may be measured in different ways. While most economic and fiscal impacts can be measured in monetary terms, assessments of social and enviromental impacts will be difficult to quantify. Because measurements of each of these sets of impacts are not made in common units, they cannot be combined in a single number to represent net program impacts.

The most feasible way to treat this problem is to list the various impacts, some with quantitative measures and some with descriptive measures, and leave the problem of combining them to final decision makers. For this approach to be satisfactory, however, primary emphasis must be placed on identifying program impacts on protecting drinking water quality. Other, secondary effects should then be identified with the remaining resources and staff time that can be devoted to evaluation, with potential unintended adverse impacts being given the most attention.

Attempts to determine what would have occurred in the watershed without the management program is a second problem that will arise. This problem involves determining the difference in conditions with a program as compared to conditions without a program. Hypothetical conditions without the program must be estimated using statistical manipulation based on comparisons with other communities or past trends within the locality which adopted the program. Since there are many variations between communities, the use of past trends in the same community is probably the most appropriate in the measurement of impacts for its watershed management program.

Another issue is how to treat time in the impact evaluation. Program effects may be measured either cumulatively or at a single point in time. Since program implementation and enforcement as well as water quality will vary significantly through time, an impact assessment should focus on cumulative differences whenever possible. Trends towards improved water quality, for example, are much more meaningful than static figures that give no indication of whether conditions are improving or getting worse.

Using Impact Assessments

In conclusion, probably the most important consideration in measuring impacts of watershed management is to determine if protection of public drinking water supplies is being achieved. Once this measurement has been made, other considerations have to be taken into account in analyzing evaluation findings. Even when a watershed management program is properly implemented and yielding positive results, the evaluator must still determine:

1. How large a positive effect is needed? At what point is water quality adequately protected?

2. How generalizable is the outcome of the evaluation? Will the positive results obtained from a watershed management program continue in the future?

3. How policy relevant are the findings? Is the information provided good enough for making decisions about committing resources on a long-term basis?

Answers to these questions depend upon local conditions,

the objectives of a community's management program, and
evaluation methods that are used. The value of an evaluation
may be undermined if these questions are not given some
consideration prior to using its findings. Decision makers,
for instance, may discredit the validity of an evaluation if
answers to these questions are not available.

COST-EFFECTIVENESS ANALYSIS

As described in the previous chapter, cost-effectiveness
analysis allows comparisons and rank-ordering of programs in
terms of achieving program objectives. In addition to being
a valuable tool for designing watershed management programs,
cost-effectiveness analysis can be used to redirect management efforts in more effective ways. The key to using cost-
effectiveness evaluation for revising management initiatives
is that it allows a comparison between existing management
methods and those that are proposed for protecting water
supplies.

Once the procedures explained in Chapter VI are used to
compare the costs and impacts of the actual and proposed
management programs, a decision can be made about whether to
reform water supply protection measures. This judgment is
made based on local political, economic, environmental, and
social conditions. As with designing a management program,
there is no set procedure for making this determination but
local officials will have an easier time assessing the worth
of reforms if costs and impacts are accurately accounted.

In summary, cost-effectiveness analysis may be viewed as
a step that is added to impact assessment evaluations. The
only major difference between the two is that with cost-
effectiveness analysis the impacts of two or more programs
are determined and compared while with impact assessments
generally only one program is evaluated. This means that the
evaluator must exercise much more judgment in making
predictions about programs' costs and impacts, especially
since some of the programs being evaluated exist only on
paper.

SUMMARY

Watershed management evaluations are technical and
political activities that can influence the decision-making
process if they are properly conducted. Whatever type of
evaluation is used--program monitoring, impact assessment,
or cost-effectiveness analysis--they are part of a continuing

process of design, implementation and reform that must be maintained if water supply protection activities are to remain up-to-date and effective.

In addition to being part of a continuous process, watershed management evaluations should be as systematic and creditable as possible if they are to overcome the wide variety of obstacles associated with any investigation of water quality and land use relationships. The procedures outlined in this chapter for conducting evaluations should assist in assessing local programs. However, as with most tasks, there is much room for inventiveness among the people closest to the programs, conditions, and constraints that are being evaluated.

REFERENCES

1. Rossi, Peter H., Howard E. Freeman, and Sonia R. Wright. Evaluation: A Systematic Approach, Sage Publications, Inc., Beverly Hills, Calif., 1979.

2. Miller, Todd L. and Raymond J. Burby with Edward J. Kaiser and David H. Moreau. Protecting Drinking Water Supplies Through Watershed Management: A Casebook for Devising Local Programs, Center for Urban and Regional Studies, The University of North Carolina at Chapel Hill, Chapel Hill, N.C., August 1981.

3. Lichfield, Nathaniel, Peter Kettle, and Michael Whitbread. Evaluation in the Planning Process, Pergamon Press, New York, 1975.

4. Hatry, Harry P., Louis H. Blair, Donald M. Fisk, John M. Breiner, John R. Hall, Jr., and Philip S. Schaenman. How Effective Are Your Community Services? Procedures for Monitoring the Effectiveness of Municipal Services, The Urban Institute, Washington, D.C., 1977.

5. Finsterbusch, Kurt. Understanding Social Impacts, Assessing the Effects of Public Projects, Sage Library of Social Research, Sage Publications, Beverly Hills, Calif., 1980.

6. Hatry, Harry P., Louis H. Blair, Donald M. Fisk, John M. Breiner, John R. Hall, Jr., and Philip S. Schaenman. How Effective Are Your Community Services? Procedures for Monitoring the Effectiveness of Municipal Services, The Urban Institute, Washington, D.C., 1977.

7. Rossi, Peter H., Howard E. Freeman, and Sonia R. Wright. Evaluation: A Systematic Approach, Sage Publications, Inc., Beverly Hills, Calif., 1979.

8. Poister, Theodore H., James C. McDavid, and Anne Hoagland Magoun. Applied Program Evaluation in Local Government, D.C. Heath and Company, Lexington, Mass., 1979.

9. Better Monitoring Techniques Are Needed to Assess the Quality of Rivers and Streams: Vol. I, U.S. General Accounting Office, CED-81-30, Washington, D.C., April 30, 1981.

10. Rossi, Peter H., Howard E. Freeman, and Sonia R. Wright, Evaluation: A Systematic Approach, Sage Publications, Inc., Beverly Hills, Calif., 1979.

11. Mace, Ruth. Municipal Cost-Revenue Research in the United States, Institute of Government, The University of North Carolina at Chapel Hill, 1961.

12. Schaenman, Philip S. Using an Impact Measurement System to Evaluate Land Development, The Urban Institute, Washington, D.C., 1976.

13. Christensen, Kathleen. Social Impacts of Land Development, The Urban Institute, Washington, D.C., 1976.

14. Keyes, Dale L. Land Development and the Natural Environment: Estimating Impacts, The Urban Institute, Washington, D.C., 1975.

15. Muller, Thomas. Economic Impacts of Land Development: Employment, Housing, and Property Values, The Urban Institute, Washington, D.C., 1976.

16. Burchell, Robert W. and David Listokin et al. The Fiscal Impact Handbook, The Center for Urban Policy Research, New Brunswick, N.J., 1978.

17. Warner, Maurice L. and Edward Preston. A Review of Environmental Impact Assessment Methodologies, prepared for the U.S. Environmental Protection Agency, U.S. Government Printing Office, Washington, D.C., 1974.

18. Herr, Philip, Gene Slater, and Robert Bluhm. Evaluating Development Impacts, Environmental Impact Assessment Project, Laboratory of Architecture and Planning,

Massachusetts Institute of Technology, Cambridge, Mass., Revised October 1978.

19. Better Monitoring Techniques Are Needed to Assess the Quality of Rivers and Streams: Vol. I, U.S. General Accounting Office, CED–81–30, Washington, D.C., April 30, 1981.

20. Miller, Todd L. and Raymond J. Burby with Edward J. Kaiser and David H. Moreau. Protecting Drinking Water Supplies Through Watershed Management: A Casebook for Devising Local Programs, Center for Urban and Regional Studies, The University of North Carolina at Chapel Hill, Chapel Hill, N.C., August 1981.

21. U.S. Department of Housing and Urban Development. The Fiscal Impact Guidebook, HUD USER, P.O. Box 280, Germantown, Md., 1979.

22. U.S. Department of Housing and Urban Development. A Practitioner's Guide to Fiscal Impact Analysis, HUD USER, P.O. Box 280, Germantown, Md., 1980.

23. Miller, Todd L. and Raymond J. Burby with Edward J. Kaiser and David H. Moreau. Protecting Drinking Water Supplies Through Watershed Management: A Casebook for Devising Local Programs, Center for Urban and Regional Studies, The University of North Carolina at Chapel Hill, Chapel Hill, N.C., August 1981.

256

Chloroform 3, 4
Cholera 14, 15
Christina Basin (Delaware), 21
Chromium 3
Cincinnati, Ohio 18
Cities and municipalities: acquire land 179; and critical areas 69; and land use 50-52, 168, 213; and taxation, 184; and water pollution 21, 38; and watersheds, water systems, and water quality 1, 7, 15, 20, 30, 38, 39, 40-42, 48, 56, 123, 185, 198, 212-213, 215; and zoning, see Zoning, regulations, ordinances, and techniques; land and water resource plans for 217; operate sewage systems 20, 40, 215; operate wastewater treatment plants 38, 39, 40-42, 56; regulate subdivision 68; waste discharges of 178; water needs of 113. See also Local governments; Urbanization; and specific cities and municipalities
City of West Frankford v. Fullop, infra. 212-13
Clays 2, 3
Climatic events and factors 6, 9
Cluster and Planned Unit Developments 123
Cluster zoning. See Zoning regulations, ordinances, and techniques
Coagulation 10, 16
Coastal Plain 151
Cobalt 3
Cochituate Reservoir (Mass.) 15
Coliform bacteria 9
Colorado 32, 34
Commercial development 9,

10, 47, 89, 97, 128, 131, 135, 137, 139, 167, 170, 171, 176, 182, 208, 236, 245
Computers and computerization, 98, 99, 101, 106, 199
Congress, U.S. 18, 19, 21, 207, 215. See also United States government
Connecticut 32, 34, 41
Constitutions: state 180, 207, 215, 216; United States 168, 180
Construction activities 9, 38, 39-40, 48-50, 98, 122, 123, 124, 129, 131, 134, 141, 144-45, 146, 169, 175, 176, 177, 178, 199, 203, 204, 209, 238, 247
Contaminants and contamination. See Pollution, water
Copper 3, 9
Copper sulfate 134
Council on Environmental Quality 88
Counties: and Best Management Practices 151; and critical areas 69; and land use 168; and roads 206; and soil conservation 206, 217; and taxes 184; and watersheds, water systems, and water quality 30-33, 38, 42-43, 48, 49, 50, 52, 54, 56, 61, 70, 72, 98, 198; and zoning, see Zoning regulations, ordinances, and techniques; extension services of 187; regulate subdivisions 68. See also Local governments; Urbanization; and specific counties
Courts. See Legal actions, cases, factors, and constraints
Critical areas 51, 55, 58, 59, 60, 69, 70, 71, 125, 139, 140, 153, 156, 157, 158, 159, 169, 176, 181
Crops and cropland. See Agriculture

258

206; demand for 204; erosion of, see Soil: erosion of; limited supply of 98; market for 122; Resource Areas for 151, 156, 157, 158, 159; resources of 98; surface models for runoff on 99-107; use and development of, see Land use and land development; value of 55, 61, 168, 179, 181, 203, 204

Land fills 137

Landowners. See Land use and land development

Landscaping 141, 144, 146, 151

Land use and land development: along reservoirs, 204; and agriculture 88, 131, 138, 139; and Best Management Practices, see Best Management Practices; and cities and municipalities 50-52, 168, 213; and construction, see Construction activities; and counties 168; and environment 228; and health 206, 208, 212; and housing market 244; and industry 131; and legal actions 206; and local governments 19-22, 68-69, 81, 135, 170; and pollution 4, 6, 7-12, 13, 15, 19, 20, 21, 29, 46, 47, 50, 130; and property rights 248; and rural areas 168; and sewer systems 183; and soil and soil conservation 123, 217; and states 19, 20, 68-69; and statutes 208; and stream quality 62; and target sources 131; and taxation 183-84; and urbanization 1, 8, 29, 46, 54;

and U.S. Constitution 168, 209; and watersheds, water systems, and water quality 4, 11-12, 21, 43, 44, 50, 52, 53, 61-62, 81, 88, 90, 107, 111, 114, 118, 120, 122, 123, 129-30, 131, 133, 135, 136, 139, 144-45, 188, 197, 199, 202, 206, 207, 208, 212, 219, 221, 223, 226, 227, 228, 236, 238, 240, 241, 243, 245, 246, 247; commercial, see Commercial development; controlled and regulated 15, 19, 31, 51, 54, 68, 83, 135, 144-45, 166, 168, 170, 174-75, 182, 183, 206, 207, 208, 209-10, 211, 212, 213, 215, 219, 220, 221, 226, 241, 242 (and see Zoning regulations, ordinances, and techniques); costs of 244; data on 62; decisions affecting 246; discouraged 51, hazardous 55, 205, 227; industrial, see Industry; in specific areas, 122, 208; management of 19; measured 245; patterns of 98; plans, classifications, and management strategies for 86, 88, 93, 137-40, 165, 183, 198, 217, 228; problems in 52; residential, see Residential development; revenue generated by 244; scientific understanding of 240. See also specific types of land use and development

Laundries and laundromats 131

Lawrence, Mass. 16

Laws. See Legal actions, cases, factors, and constraints; Legislation

Lead 3, 9

Legal actions, cases, factors, and constraints 53, 54, 60, 62, 69, 117, 120, 133, 166-68, 169, 184, 186, 187-88,

261

195, 200, 205, 206, 207-
13, 216, 228, 236, 246.
See also Legislation
Legislation: state 68-69,
71, 118, 168, 179, 184,
206, 210, 211, 212, 215,
216, 223; United States,
see United States Congress;
various 133, 214, 225.
See also specific legis-
lative acts
Litter 130, 134, 148, 160,
185
Livestock 8, 39, 57, 147.
See also Agriculture;
Animals and animal waste
Local governments: acquire
land 179, 181, 182; and
easements 181; and land
use 19-22, 68-69, 81,
135, 170; and states 211;
and taxation 184; and
urbanization, see Urbani-
zation; and watersheds,
water systems, and water
quality 17, 19-22, 29-75,
81, 166, 167, 168, 175-76,
181, 182, 186-87, 188,
195, 198, 205, 206, 207-
13, 214, 215, 216, 217,
228, 238, 240, 246-47,
249; and zoning, see Zon-
ing regulations, ordi-
nances, and techniques;
authority and power of
184, 211; control public
services, 183; future
problems of 42-43. See
also Cities and municipalities; Counties
Loch Raven Reservoir (Md.)
245
Logging 38
London, England 14, 16
Long Pond (Mass.) 15
Louisiana 18, 21, 32, 34
Louisville, Ky. 16

Maine 32, 34, 161, 177,
178, 187, 188, 205, 206

Magnesium 3
Major Land Resource Areas
(MLRAs) 151, 156, 157,
158, 159
Management of watersheds:
advantages and benefits of
62, 63-65, 117; agencies
and governments implementing
31, 43-50, 237, 238, 241
(and see Cities and municipalities; Counties; Local
governments; Regions;
States); alternatives in
62, 117, 119, 126, 150, 195,
201-4, 219, 220, 222, 228;
and agriculture 121; and
constituent groups 222,
223, 224; and income re-
distribution 248; and land
acquisition, see Land:
acquisition of; and land use
and development, see Land
use and land development;
and local officials, 223;
and pollution, see Pollu-
tion, water; and soil conservation 206; and state
laws 69; and water quality,
see Water quality; and water
systems 205-6; backbone of
166; basic directions and
dimensions of 111-63; capital improvements program in
221; citizen task forces for
227; controls in 81, 135-50,
165, 228; costs, cost-effectiveness, funding, and fiscal
impacts of 48, 65, 116, 117,
118, 119, 122, 151, 155, 156-
59, 166, 195, 198-204, 219
220, 221, 222, 224, 227, 228,
235, 236, 238, 239, 240-41,
243-44, 246, 249, 251; decision guides in 165; design
and implementation of programs in 82, 111, 115, 118,
119-26, 127, 134, 189, 195-
233, 235, 236, 242, 243, 247,
252; direction and goal setting in 81, 82, 111-63,

262

Oklahoma 32, 34
OMB (U.S.) 214
Open spaces. See Parks and open spaces
Orchards 152, 156
Ordinances: on erosion control 50, 58; on land use 183, 209; on sedimentation 58; on vegetation 124; on watersheds and water quality 51, 55, 58, 59, 60, 63, 69, 74, 133, 168, 176, 182-83, 209, 210, 212, 228, 245; on zoning, see Zoning regulations, ordinances, and techniques
Oregon 32, 34
Organic chemicals and materials 2, 3-4, 7-8, 11, 128. See also chemicals
Oxygen and oxygen-demanding materials 33, 37, 104, 106

Pacific Northwest (U.S.) 31, 32, 39
Package sewage treatment plants 10
Paper industry 131
Parks and open spaces 50, 66, 67, 114, 123, 124, 139, 141, 171, 175, 184, 246, 248. See also Recreation and recreational areas
Particulates 2-3, 6, 128
Pasture land. See Agriculture
Pathogenic organisms 2, 16
Pee Dee River basin 218
Penn Central Transportation Company v. City of New York 209
Pennsylvania 32, 34, 61-62
Pequannock Watershed (N.J.) 15, 196, 208
Personnel, professional and technical 53, 54, 60, 61, 120, 196-98, 199, 200, 203, 221, 224 241

Pesticides 3, 4, 8, 9, 38-39, 40, 69, 70, 71, 102, 130, 140, 155, 203
Pesticide Transport and Runoff Model 102
Petroleum industry 33, 131, 173
Phenolics 15
Phosphorus 3, 4, 100, 106, 121, 134
Photosynthetic production 106
PHS. See United States Public Health Service
pH value 3
Piedmont 122-25, 151
Pipelines 10, 89
Planned Unit Development (PUD) 172-73. See also Zoning regulations, ordinances, and techniques
Plants. See Vegetation
Plumbing codes 177
Point source permits 69, 70
Point sources of water pollution 7, 14, 21, 38, 39, 51, 56-57, 69, 70, 132, 178, 215. See also Source protection of water and watersheds; and specific point sources of pollution
Poisoning 18
Politics and political factors 53, 54, 60-61, 63, 98, 112, 114, 117-18, 119, 120, 166, 175, 179, 183, 195, 196, 198, 200, 216-19, 220, 222, 223, 224, 226, 227, 228, 240, 241, 251
Pollution, air 40
Pollution, water: and accidents 7; and Best Management Practices 155; and employment levels 132; and forests 39; and health 1, 2, 3, 4, 5, 11-22, 29, 37, 38, 44, 46, 50, 52, 65, 82, 83, 85, 88, 90, 93, 96, 121, 122, 128, 132, 166, 168, 169, 178, 226, 245, 248; and impervious surfaces 132; and land surface models

265

100, 101; and land use 4, 6, 7-12, 13, 15, 19, 20, 21, 29, 46, 47, 50, 130; and population 132; and private landowners 47; and private secotr 48-50; and raw water 1, 19; and urbanization 33, 38, 40-42, 48; and wastewater 7; and water systems and watersheds 33, 35, 56-57, 63, 66, 67; and zoning 52; by source, matrix for 90-91; legislation on 69; sources, nature, types, pathways, control and effects of 1-28, 29, 38, 39, 40-42, 44, 46-47, 48, 50, 56, 57, 62, 66, 67, 69-70, 75, 81-107 passim, 112, 115, 119, 123, 126-32, 133, 134, 136, 137, 144, 148, 149, 150, 151, 160-61, 165, 171, 172, 173, 174, 176-77, 178, 183, 185, 186, 199, 202, 206, 208, 211, 213, 215, 218, 219, 220, 221, 223, 236, 244, 245, 247, 248. See also Nonpoint water pollution; Point sources of water pollution

Ponds 31, 32, 48, 49, 73, 128, 149, 154, 157, 158

Population 16, 20, 33, 85, 89, 132, 171, 172, 204, 208, 220

Portland (Maine) Water District 161, 178, 187, 188, 205, 206

Potomac River Basin 8

Poughkeepsie, N.Y. 16

Precipitation. See Rainfall and precipitation

Preferential taxation. See Taxes and taxation

President's Council on Environmental Quality 18

Press 222, 223, 227, 228

Private property owners. See Land use and land development

Property rights, purchase of. See Land: acquisition of

Property taxes. See Taxes and taxation

Protozoa 2

Public health. See Health

Public Health Service, U.S. 13, 14, 17

Public Law 74-461 218

Public Law 92-500 21, 33, 65-67, 187, 213, 215

Public Law 93-523 1, 12, 18-19, 21, 215-16

Public Law 566 217, 218

Public Lae 703 217

PUD 172-73. See also Zoning regulations, ordinances, and techniques

Pulp and paper industry 131

Radio 222, 227

Radioactivity and radioactive materials 2, 4, 8, 128, 140

Railroads 10, 89, 131. See also Transportation facilities and systems

Rainfall and precipitation 5, 6, 7, 9, 10, 40, 86, 89, 101, 124

Raw water 1, 11, 14, 15, 16, 17, 19, 20, 22, 29, 31, 35-38, 40-43, 44, 45, 46, 56, 63-64, 65, 74, 111, 112, 113, 115, 116, 117, 131, 132, 135, 140, 172, 183. See also Pollution, water; and specific types of impoundments

Real estate development. See Land use and land development

Recreation and recreational areas 9, 17, 44, 48, 66, 67, 97, 114, 116, 122, 131, 139, 140, 141, 142-43, 154, 158, 159, 173, 174, 175, 178, 217, 223, 224, 236, 248. See also Parks and

open spaces
Refuse disposal sites 140
Regional councils of government. See Regions
Regions: and land use 68; growth of, and watersheds 88; of United States, specified 31; of United States (I-X), statistics on 31, 32, 33, 34, 39, 41; regulate water quality and watersheds 21, 22, 29-75, 99, 121, 205, 212, 213-14, 215, 216, 238. See also specific regions
Regression analysis 106
Reservoirs and impoundments: and chemicals 161; and environment 139; and pollution 6, 8, 9, 20, 83; and raw water 140; and sewers 113; and special districts 174; and streams 45; and water quality 104-5, 113; and water systems 17; as critical areas 139; as drinking water source 32; as raw water source 31; buffer strips around 45, 215; eutrophication of 21, 106; land acquired for, see Land: acquisition of; land use along, 204; loss of storage in 64, 203; media events at 228; phosphorus in 100; plans for 140; recreation at 17, 142-43, regulated 140, 167, 178; sedimentation in 245; specific 15, 134, 235, 245; streams feeding, see Rivers, streams, and water courses; treated 135
Reservoir Water Quality Model 105
Residential development 7, 9, 10, 47, 48, 51, 54, 55, 58, 59, 63, 68, 89, 97,

98, 123, 129, 131, 135, 137, 138, 139, 149, 167, 169, 170-71, 172, 174-75, 182, 183, 186, 202, 203, 204, 208, 215, 236, 244, 245, 247, 248
Resource Conservation and Development Program 217-18
Resource Conservation and Recovery Act 19
Resources Conservation Act 218
Restaurants 131
Rhode Island 32, 34
Rivanna Water and Sewer Authority 196, 212
River Basin Survey Program 218
Rivers, streams, and water courses: and chemical treatment 161; and environment 128, 139; and hydrology 134; and land use 62; and livestock 147; and lot size 171; and pollution 6, 13, 20, 38, 40, 56, 83, 86, 93, 96, 134, 178; and recreation 173, 175; and reservoirs 45; and septic tanks 52; and sewers 160; and water quality 104, 105, 132; and watersheds 104, 217; as water source 31, 32, 56; basins of 217, 218; buffer strips along 45, 46, 139, 140, 141, 175, 199, 203, 207; channelization of 124; chlorinated 161; corridors of 61-62; erosion along 40, 125; flow of 6, 40, 44, 89, 97, 134; illegal discharges into 130; maps of 139-40; protecting banks of 144; quality of 62; sedimentation in 8; soil along 139, 141; special districts for 174; stabilization of 147, 154, 156, 157, 158, 159; topography along 139; treated 135;

268

269

(HUD) 72, 247
United States Environmental
Protection Agency (EPA)
11, 14, 18-19, 21, 30, 41,
43-44, 65, 72, 100, 132,
156, 157, 158, 159, 201.
See also Environment, en-
vironmental protection,
and environmental impact
statements
United States Farmers Home
Administration 72
United States General
Accounting Office 62
United States Geological
Survey 72
United States government:
and evaluations 235;
and river basin planning
218; offers tax incen-
tives 185; regulates
environment 244; regu-
lates land use 19; regu-
lates water quality and
watersheds 18, 19, 20,
22, 29, 31, 53, 54, 60,
61, 62, 65, 68, 71-74,
119, 121, 166, 169, 186-
87, 205, 206, 210, 213-19.
See also United States
Congress; and specific
governmental agencies
United States Office of
Management and Budget
(OMB) 214
United States Public Health
Service (PHS) 13, 14, 17
United States Soil Conservation Service (SCS) 47,
71-73, 121, 150, 151,
187, 216, 217-18
United States Supreme Court
209-10
United States Treasury Department 17
Universal Soil Loss equation
101
Urbanization: and agricul-
ture 184; and Best
Management Practices 47

73, 150; and Boston 15;
and environment 139, 248;
and land acquisition 180;
and land use 1, 8, 29, 46,
54; and local governments
50; and on-lot disposal systems 177; and raw water
37; and sewer systems 20;
and Soil Conservation Service 73, 217; and water
flow 100; and water
quality and water pollution
8-10, 15, 18, 20, 22, 29,
38, 39, 40-42, 43, 47, 50,
55, 71, 73, 133, 185; and
watersheds and water systems
1, 16, 22, 33-43, 46, 47,
54, 55, 56, 61, 74, 101,
114, 115, 123, 130, 131,
149, 150, 172, 182, 184,
185, 246; and water supply
1; controlled 135, 138,
139, 145-50; increases 1,
29; in specific areas 182,
208; in suburbs 130, 208;
plans for 138, 139; rain-
fall in areas of 7. See
also Land use and land
development
USDA 38, 215, 218
Utah 32, 34

Vanadium 3
Vegetation 2, 9, 39, 40, 48,
49, 89, 123, 124, 125, 146,
147, 148, 152, 153, 154,
155, 156, 157, 158, 159,
175, 176, 178
Vehicles. See Motor Vehicles;
Trucks and trucking industry
Vermont 32, 34
Veterinarians 137
Vineyards 152, 156
Vinyl chloride 3
Virginia 32, 34, 134, 161,
170, 173, 176-77, 180, 183,
187, 196, 198, 212, 213,
227-28, 235-36
Virginia Water Pollution Con-
trol Board 213

access to 15, 122; regu-
lated 68, 122, 207, 208,
209, 216, 229; sewer lines
in 58, 59; sizes of 6,
53, 120; small program for
217; soil loss in 203;
water exported from 198;
water lines in 51, 52,
58, 59; water uses in
107; water yield and quality in 29, 63, 64, 118.
See also specific topics
"Water Supply and Ground
Water Considerations in
Preparation of a 208
Wastewater Management
Plan" 65
Water supply systems: ac-
quire land, see Land:
acquisition of; age of
47; and domestic water
use 7; and local govern-
ments 44, 45, 46, 47;
and other governmental
agencies 224; and raw
water 35-37, 43; and
seasonal water shortages
35, 37, 42-43, 63, 64; and
source protection 224;
and state legislation 68-
69; and technology 1; and
wastewater 20; and water
quality 22; and water-
sheds, see Management of
watersheds and Watersheds;
and water yield 35, 37;
community 17; control
flow 48, 49; costs and
financing of 11, 246;
deficiencies in 18; ex-
ecutives of 30; future
demands on and problems
of 29-75; import water
31-32; in specific places
198, 206; management ex-
perience and techniques in
29-75; median, in United
States 31; nature and
responsibilities of 205-
6; need additional water

35; number of 1, 17, 30;
ownership of, jurisdiction
over, and types of 1, 7,
11, 17, 18, 19, 30-31, 228;
raw water sources of 31;
regulated 18; revenues of
47; size of 31, 44, 47;
standards of 19; surveyed
and monitored 17, 29-75;
technology of 1, 11; treatment by, see Water treatment
and water treatment plants;
wastewater from 21. See
also specific topics
Water treatment and water
treatment plants 1, 2, 3,
4, 10-11, 13, 14, 15-22, 29,
37, 40-43, 47, 51, 65, 83,
85, 86, 87, 92, 115, 134,
136, 138, 173, 215, 245,
246
Weather 155
West Milford, N.J. 170
West Virginia 32, 34
Wetlands 89, 124, 139, 248
Wildlife 44, 114, 155, 159,
217, 218
Wind erosion 6, 9, 154, 155,
156, 158, 159
Wisconsin 32, 34, 39
Woods and woodlands. See
Forests and forest management
Wyoming 32, 34

Yadkin-Pee Dee river basin,
218

Zinc 3, 9
Zoning regulations, ordinances,
and techniques 51, 52, 54,
55, 56, 58, 59, 68, 123, 124,
129, 136, 167, 169, 170-74,
181, 183, 199, 202, 208, 210,
211-12, 215, 216, 220, 226,
236, 242, 248
Zooplankton 106